The Political Ecology of the State

T0199737

The contemporary state is not only the main force behind environmental change, but the reactions to environmental problems have played a crucial role in the modernisation of the state apparatus, especially because of its mediatory role.

The Political Ecology of the State is the first book to critically assess the philosophical basis of environmental statehood and regulation, addressing the emergence and evolution of environmental regulation from the early twentieth century to the more recent phase of ecological modernisation and the neoliberalisation of nature. The state is understood as the result of permanent socionatural interactions and multiple forms of contestation, from a critical politico-ecological approach. This book examines the tension between pro- and anti-commons tendencies that have permeated the organisation and failures of the environmental responses put forward by the state. It provides a reinterpretation of the achievements and failures of mainstream environmental policies and regulation, and offers a review of the main philosophical influences behind different periods of environmental statehood and regulation. It sets out an agenda for going beyond conventional state regulation and grassroots dealings with the state, and as such redefines the environmental apparatus of the state.

Antonio Augusto Rossotto Ioris is Lecturer in Environment and Society, School of Geosciences, University of Edinburgh. His main areas of research are related to the search for environmental justice in the urban and regional contexts, and the multiple obstacles faced by marginalised communities to influence environmental decision-making. He has extensive experience with policy-making and project management in the UK, Portugal, Brazil and Latin America.

Routledge Studies in Political Ecology

The *Routledge Studies in Political Ecology* series provides a forum for original, innovative and vibrant research surrounding the diverse field of political ecology. This series promotes interdisciplinary scholarly work drawing on a wide range of subject areas such as geography, anthropology, sociology, politics and environmental history. Titles within the series reflect the wealth of research being undertaken within this diverse and exciting field.

Published:

The Political Ecology of the State
The basis and the evolution of environmental statehood
Antonio Augusto Rossotto Ioris

The Political Ecology of the State

The basis and the evolution of environmental statehood

Antonio Augusto Rossotto Ioris

Routledge
Taylor & Francis Group

LONDON AND NEW YORK

First published 2014
by Routledge
2 Park Square, Milton Park, Abingdon, Oxon OX14 4RN

and by Routledge
52 Vanderbilt Avenue, New York, NY 10017

First issued in paperback 2020

Routledge is an imprint of the Taylor & Francis Group, an informa business

British Library Cataloguing in Publication Data
A catalogue record for this book is available from the British Library

Library of Congress Cataloging in Publication Data
Ioris, Antonio Augusto Rossotto.
The political ecology of the state: the basis and the evolution of
environmental statehood/Antonio Augusto Rossotto Ioris.
  pages cm
Includes bibliographical references and index.
  1. Political ecology. 2. State, The. 3. Environmental policy. I. Title.
JA75.8.I67 2014
320.1–dc23

                                                   2013036043

ISBN 13: 978-0-367-66957-7 (pbk)
ISBN 13: 978-0-415-72219-3 (hbk)

Typeset in Times New Roman
by Wearset, Boldon, Tyne and Wear

Contents

Preface

The object of this book is to reinterpret the recent, and still ongoing, adjustments in the structure and functioning of the contemporary (capitalist) state which aim to respond to environmental pressures and associated forms of criticism. The central question here is how the mobilisation of civil servants, technology and resources for environmental management issues has become one of the most active, but also frustrating, areas of policy-making and state interventions. The term environmental statehood will be used in forthcoming chapters to define the combination of discursive, ideological and material efforts by the state to deal with socioecological problems. The different forms of environmental statehood contain specific institutional mechanisms and rhetorical devices that, despite the best intentions, are not always consistent, or necessarily coordinated. The formation and operation of environmental statehood actually encapsulate manifold disputes fought over values, assumptions and rationalisations. The introduction of a given form of environmental statehood represents, at best, a tentative reaction to ecological disruption, political tensions and economic demands. This suggests a strong case for reconciliation between work on state theory and ecological politics.

The political and operational complexity of the contemporary state needs to be carefully examined taking into account its socioecological 'signature'. Instead of an entity detached from nature and society, the state is a main politico-ecological player deeply embedded in socionatural relations and political conflicts. Interestingly, the environmental responses articulated through the organisation of environmental statehood not only affect socioecological systems, but also influence the basis of statecraft. In recent decades, the globalisation of society and the markets has led to the modification of statecraft processes and rescaling of environmental statehood. The environmental agendas of the state now spread from the traditional responsibilities at national level to both the local and international (or multilateral) spheres of public administration. This evolution of statehood has happened not only through changes in the configuration of the state, but includes qualitative alterations related to the wider reforms of the state apparatus. In the last 100 years or so, environmental statehood has evolved from narrow and bureaucratised approaches into sophisticated strategies of public engagement and self-regulation. The latter have been epitomised by the expression environmental 'governance' to replace the more conventional forms of environmental 'government'. The main dilemma faced by politicians and academics is that these different models of environmental statehood – formulated respectively under liberal, Keynesian and neoliberal economic theories – have been able to produce only limited solutions to intricate environmental management problems.

A politico-ecological perspective is, therefore, essential to understanding the socioecological sensitivities and constraints of the state. The central claim of this book is that the controversies affecting environmental statehood lie, primarily, in its historical role and class commitments. The contemporary state – created in Europe during the consolidation of industrial capitalism and then disseminated to the rest of the world – has among its core functions the advance of the anti-common strategies needed for the maintenance of prevailing production and accumulation processes. This fundamentally anti-ecological feature of the state pervades not only its economic and social interventions, but also the formulation of legislation and regulatory instruments apparently aimed at ecological restoration and conservation. This means that environmental statehood is inherently and systematically contradictory or, as argued in the following pages, characterised by a rational irrationality that has its roots in the ideological and politico-economic foundations of the state. The practice of environmental statehood demonstrates that it is chiefly intended, through environmental protection discourses and formally restrictive regulation, to stabilise the mechanisms for the expansion of capitalist relations and to foster new avenues for the circulation of capital. The constant adjustments to environmental statehood throughout the last century are nothing less than an adjunct to the revitalisation of capitalist production and reproduction.

The origins and reorganisation of the mainstream models of environmental statehood have been directly and indirectly informed by old and new theories about politics, property and the state. Three authors seem to have been the main sources of inspiration behind the trajectory of environmental statehood, namely, Hobbes, Kant and Hegel. The latter's political philosophy is specifically associated with the upholding of the flexible model of environmental statehood associated with the rhetoric and tactics of governance. However, the claims of rationality, legitimacy and freedom made under the influence of Hegelian theories have been frontally contradicted by the ineffectiveness of actual state interventions and the perpetuation of environmental impacts, socioeconomic discrimination and political inequality. Probably some of the most emblematic examples of the failures of environmental statehood are the pervasiveness of carbon-based lifestyles, the use of private cars and mass consumption of goods. The consequence is the impending threats of climate change and massive socioecological disruption, while the state remains powerless and unable to move away from inertial, and essentially suicidal, trends (an example is the collapse of the negotiations on climate change mitigation in Durban in 2011 and Doha in 2012).

On a more personal note, it should be explained that this modest book is the result of more than ten years of experience in environmental policy-making and academic research (primarily, but not only, related to water allocation, use and conservation). The intense experiences and varied opportunities to learn about the arrangement and the functioning of environmental statehood have highlighted the existence of this unresolved paradox: agencies and legislation are growing in sophistication while environmental problems also keep expanding in space, scale and complexity. There is a structural difficulty in resolving environmental problems and in eliminating social injustices related to environmental management. After years dedicated to thinking and reflecting upon this paradox, I came to the conclusion that the more the state deals with environmental issues, the more it promotes or endorses the exploitation of socionature and widens the gap between society and its ecological condition. This is the raw material and the starting point of this investigation.

It is quite possible that my analysis may fail to offer a completely satisfying assessment of these issues, but hopefully it can be seen as a useful exercise in critical thinking and a contribution to the politico-ecological debate. As will become noticeable very soon, the current work endeavours to be faithful to the radical left-wing tradition, where I believe that significant and highly relevant assistance can be found for dealing with the achievements and shortcomings of the contemporary (capitalist) state. Finally, I wish to mention that this book is dedicated to my little son, Antonio, who was born during the preparation of the manuscript and just before my arrival at the University of Edinburgh in the middle of 2012. The fathering experience added even more sleepless hours to an already demanding writing process, but also provided me with an extraordinary amount of passion, inspiration and optimism about the possibility of seeing meaningful changes in the world. The book is also lovingly dedicated to Adriane, who brought our beautiful son to this world and has been the most wonderful company and support for many years.

Earlier versions of the sections of some chapters originally appeared as:

Ioris, A.A.R. 2008. Water Institutional Reforms in Scotland: Contested Objectives and Hidden Disputes. *Water Alternatives*, 1(2), 253–70 [in Chapter 5].

Ioris, A.A.R. 2010. The Political Nexus between Water and Economics in Brazil: A Critique of Recent Policy Reforms. *Review of Radical Political Economics*, 42(2), 231–50 [in Chapter 4].

Ioris, A.A.R. 2012. Applying the Strategic-Relational Approach to Urban Political Ecology: The Water Management Problems of the Baixada Fluminense, Rio de Janeiro, Brazil. *Antipode*, 44(1), 122–50 [in Chapters 2 and 3].

Ioris, A.A.R. 2012. The Neoliberalization of Water in Lima, Peru. *Political Geography*, 31(5), 266–78 [in Chapter 5].

Ioris, A.A.R. 2013. The Adaptive Nature of the Neoliberal State and the State-led Neoliberalisation of Nature: Unpacking the Political Economy of Water in Lima, Peru. *New Political Economy*, 18(6), 912–38 [in Chapter 5].

1   Introduction

This book will examine the socioecological commitments of the contemporary (capitalist) state and discuss the political ideas underpinning, and often constraining, environmental statehood. Because of its many responsibilities and multiple interventions, the state is a key socioecological or socionatural player – considering the hybrid ontology of the world, simultaneously 'social' and 'natural' – and its environmental agenda helps to shape the wider public sector and also contribute to either challenging or legitimising state institutions.[1] Environmental statehood, in turn, is more than just the public administration of resources and ecosystems, but comprises the application of specific discourses, strategies and techniques. Through the advance of environmental statehood, the contemporary state has become both a mediator of socioecological conflicts and a driver of additional environmental change. Moreover, although socioecological issues have meant a great deal for the reconfiguration of the contemporary state[2], there is still a need for concerted scholarly work on the synergies between the responses to environmental problems and the underlying politico-ideological factors that influence the effectiveness of these same responses. A proper consideration of the socioecological repercussions of the contemporary state requires a reinterpretation of political theory in a way that posits environmental politics inside, and in relation to, statecraft and public policy-making.

A careful examination of the politics of environmental statehood is ever more necessary nowadays if we take into account that mainstream public policies are aimed not merely at restraining and reverting environmental degradation, but also at justifying and reinvigorating prevailing socioeconomic trends. This is evident, for example, in the case of land use management, energy generation, large engineering constructions and the exploration of biodiversity and mineral reserves sponsored or authorised by the state. Early forms of state regulation on environmental issues, such as water control and land management, are undoubtedly as old as history and form a part of the achievements of the ancient civilisations, such as in China, Egypt, Peru and Mesopotamia, as demonstrated by their sophisticated practices of river engineering, plant domestication and irrigation. Nonetheless, those processes took a more intricate role at the organisation of capitalist relations of production and reproduction. In recent centuries, the state has had to facilitate access to territorial resources and guarantee private enterprise, whilst coping with environmental impacts and mediating tensions between social groups. Our starting point in order to understand the contradictory environmental effects of the state apparatus is to consider the state as more than simply a collection of agencies and regulatory instruments. Borrowing from Lefebvre (2008), the state should be seen as complex structures and strategies that reflect the balance of political power and the growth of social antagonisms.

One of the central socioecological repercussions of the capitalist state was the creation of powerful *anti-commons institutions*, which were critical to secure the growth of commodity production and capital accumulation. Economic intensification and technological progress largely depended on the appropriation, and private exploitation, of the assets which had formerly been held collectively by serfs, peasants or indigenous tribes. Once the commons were no longer protected, the stronger and more opportunistic competitors were able to exclude others from the access to scarce, previously shared, resources.[3] But in order to do so, they needed the state to purge political opposition and safeguard private possessions. This historical phenomenon was accurately identified by Adam Smith when he conceded that:

> Wherever there is great property, there is great inequality. For one very rich man, there must be at least five hundred poor, and the affluence of the few presupposes the indigence of the many.... It is only under the shelter of the civil magistrate that the owner of that valuable property ... can sleep a single night in security. He is at all times surrounded by unknown enemies, whom, though he never provoked, he can never appease, and from whose injustice he can be protected only by the powerful arm of the civil magistrate.... *The acquisition of valuable and extensive property, therefore, necessarily requires the establishment of civil government.* Where there is no property, or at least none that exceeds the value of two or three days labour, civil government is not so necessary.
>
> (Smith, 2008: 408, emphasis added)

Following Smith's prominent socioeconomic theorisation, it is the propertied class who are at risk of suffering injustice from their 'unknown enemies' – that is, the poor – who are nonetheless the result of the affluence and the acquisition of extensive property by the rich. Instead of actually resolving this serious economic imbalance, the state should first protect the 'owners of that valuable property'. If private property has little value, then the state, as Smith argued in the middle of the eighteenth century, was pointless. Through the action of the state, anti-commons strategies became a crucial part of socioeconomic development, although this meant that mechanisms to politically contain the dissatisfaction of the poor and destitute were necessary. The material consequence of this anti-commons ideology was to leave the contemporary (capitalist) state in charge of the most decisive pressures on socioecological systems, at the same time as it also had to negotiate the rate and the distribution of negative impacts. More importantly, the anti-commons commitments of the state, essential for the success and expansion of capitalist society, are inherently antagonistic to the discourse of democracy and economic freedom that characterises the bourgeois political system. Durkheim (1957) rightly identifies this inconsistency when he explains that the aim of the state "is not to express and sum up the unreflective thought of the mass of the people but to superimpose on this unreflective thought a more considered thought, which therefore cannot be other than different" (quoted in Habermas, 1987: 81). That is certainly the case of existing environmental policies, which seem to operate in favour of the protection of the commons, but mainly in order to sustain the overarching anti-commons requirements of the mass consumption and mass waste economy.

As demonstrated by many authors – above all Marx in his work on value, commodity and politics – capitalist production is based on the private ownership of production means in the hands of non-workers. At the same time as the "capitalist mode of appropriation, which springs from the capitalist mode of production, produces capitalist private property" (Marx, 1976: 929), capitalist private property "rests on the exploitation of alien, but formally free labour" (Marx, 1976: 928). The privatisation of the commons formed the historical conditions for the transformation of socioecological processes according to the priorities of the capitalist economy. Expropriation of common land had started around the fifteenth century, but in the eighteenth century, at the dawn of the Industrial Revolution, it took the form of official "robbery" fittingly endorsed by the state, as in the notorious case of the progressive enclosures of common fields in the British countryside (Marx, 1976: 885). The anti-commons agenda of the state became more visible during this exact period of transition to intensified forms of production and workforce exploitation, during which time crimes against private property (e.g. "throwing down fences when commons were enclosed") were often paid for with capital punishments (Thompson, 1966: 60). With the advent of industrial capitalism, the previous mercantile state had to make serious adjustments, and its transformation was in itself an integral part of the expansion of capitalism in Europe and around the world. In the economic history of the Western countries, "the really important transition that needs to be elucidated is not from feudalism to capitalism but from scattered to concentrated capitalist power" (Arrighi, 1994: 11).

The management of socioecological systems and the guarantee of private property became central tasks of the contemporary state, while the strategic convergence between state and capital was necessary for the conquest of new territories and more resources. Early European state formations have been historically based on the interrelated strategies of territorialism and financial capitalism (Arrighi, 1994), while the growth of imperialism was constituted by the combination of capitalist and territorial logics of power and economic and geopolitical competition (Callinicos, 2007). The progressive expansion of the capitalist economy could not have advanced without both the encroachment upon the 'more-than-capitalist' spheres of the world, and the command exerted by the state over people and the rest of socionature. Luxemburg (1951) specifically pointed out that capitalism inevitably needs other, non-capitalist territories and people to exploit around the globe. Moreover, at the same time as the anti-commons requirements of capitalism demanded the action of the state for the continuous conversion of important features of local and global commons into opportunities for accumulation, it was also necessary that the state tacitly restrained anti-commons excesses that could affect long-term interests (such as excessive whaling in international waters, hunting of endangered species and the fast conversion of forested areas). For that reason, since the latter part of the nineteenth century, early forms of environmental degradation started to concern state officials and the public, and this concern triggered the introduction of regulation on issues such as pollution, deforestation and biodiversity loss. The effort to regulate anti-commons impulses became regular state initiatives which attempted to contain the self-destructiveness of capitalism, at least in the short term. It has been an implicit 'negation of the negation' of collectivised lands, resources and ecosystems that is vital to prolong the hegemonic processes of production and consumption.

Either of its own volition or in response to demands from pressure groups, the capitalist state had to gradually improve its handling of anti-commons strategies in order to try to reconcile economic growth with compensatory environmental measures. Through a combination of violence and persuasion, expansionism and restraint, discourse and practice, the contemporary state developed complex mechanisms to both mitigate the more urgent symptoms of environmental degradation and restrict protest and criticism. However, the search for sound environmental management by the state constitutes primarily a strategy to preserve, in time and space, economic activities that depend on the very appropriation and exploitation of the commons. Nonetheless, when considering the available literature on state theory and environmental policy-making, it is easy to detect a gap in the comprehension of the achievements, failures and possibilities of state interventions, as much as a deficit in the understanding of the actual socioecological embeddedness of state politics. In contrast to the suggestions of mainstream theorists that dominate this field of research, the green agenda of the state is never inherently conservationist and not automatically endorsed by wider society. On the contrary, the action of the state denotes values and assumptions that are integral to processes of political dispute and ideological confrontation. The socioecological complexity of the state needs to be carefully examined, taking into account its spatial, temporal and political 'signature'. The state cannot be seen as a monolithic, internally coherent entity, but rather a condensation of forces and social relations (Poulantzas, 1978).

There is, therefore, a strong case for a rapprochement between work on state theory and ecological politics. This means addressing state interventions from a politico-ecological perspective and, ultimately, crafting a political ecology framework of the state. We have to consider two different forms of environmental politics, one more implicit (consisting of policies and practices that lie outside the arenas conventionally labelled as 'environmental' where the state always played a crucial role) and the more explicit politics shaped by treaties, legislation and multiple forms of regulation (Conca, 1993). Equally, instead of something being written in stone, there is a constant production of the state through everyday practices that correspond to the dynamic 'geographies of stateness' (Painter, 2006). So far, most of the academic literature has focused overwhelmingly on the formulation and implementation of state policies, but has paid less attention to ontological changes in the state produced by disputes over socioecological demands and impacts. It has been frequently ignored that the "control and management of nature has always been central to the realization and consolidation of state power" (Whitehead *et al.*, 2007: 6). Even the examination of environmental politics – including the three main worldviews: individualism and free-market conservatism, biocentrism and deep ecology, and socialist and libertarian environmentalism (Dryzek and Lester, 1989) – rarely deals with the foundations of state dilemmas. Whereas there is extensive theorisation of concepts such as sustainable development (e.g. Dobson, 1998) and ecological modernisation (e.g. Mol and Sonnenfeld, 2000), there is only a very partial consideration of the politics that inform the socioecological responses put forward by the state.

As a result, there remains a limited conceptual understanding among political scientists of the green commitments and shortcomings of the state in the last century or so. Most of the analyses have been superficial and have fallen short of identifying the core elements of environmental statehood. For instance, failures of state policies and initiatives have been associated with the spatial disjuncture between the territorially organised state and the spatiality of ecological problems, as well as with the domination, disempowerment and violence the state tends to perpetuate (Paterson *et al.*, 2006). Paehlke and Torgerson (2005) compare the administrative state with an environmental Leviathan, and while they call for a more active involvement of social groups in state policies, the authors fail to address the underlying economic inequalities shaping the state's response to deprived groups. Eckersley (1992, 2004) insists on the importance of an ecocentric approach supporting deliberative democracy which could usher in a solution to the ecological crisis, but she too easily associates critical environmental politics (e.g. ecosocialism) with anthropocentrism and, as a result, misses the opportunity to connect the political agency of the working class with the creativity of the environmentalist movement. What is more, Bernstein (2001) identifies the misbehaviour of individuals or the lack of a proper economic treatment of ecological processes as causes of environmental problems, while Hayward (1998) argues that political theory needs to deal with the opportunities available to civil society to change policy-making and Conca (2006) examines the struggles for the institutionalisation of the 'nonstate' through elusive recommendations for integration and public participation. All the above authors stop short of establishing the deeper connections between socioecological pressures, classbased disputes and the changing patterns of the state.

In an attempt to go beyond the inadequacies of most recent publications, a pivotal argument of this book is that the main challenges affecting contemporary environmental policy-making are a reflection of other prominent debates about the configuration and politico-economic responsibilities which the state held almost 200 years ago at the rise of industrial capitalism. Now, just as then, the emergence and persistence of socioecological disputes are concentrated in the intervention and biases of the state. With the consolidation of the mass production society, one of the state's main tasks became the dealing with socioecological impacts, as well as the naturalisation and maximisation of anti-commons institutions.[4] These tasks have certainly achieved a level of unprecedented sophistication in recent years, but also replicate the earlier discussion on constitutional and institutional reforms in previous centuries. As will be demonstrated in later chapters, the achievements and limitations of today's environmental statehood can be explained particularly well with a Marxist critique of Georg Wilhelm Friedrich Hegel and his theories on idealising and legitimising a flexible, but intrinsically conservative, state rationality. It is relevant to observe that the interpretation of Hegel's political elaboration in the Anglo-American world had been influenced by idealist philosophy since the beginning of the twentieth century, but Hegel's work lost its appeal during the period of authoritarianism and Keynesian economics. It was only when the welfare-developmentalist state started to show its insurmountable contradictions that Hegelian politics, even implicitly, started to make sense again.

The renewed importance of a critique of Hegel's political ideas needs to be situated in the wider perspective of state theory. So far, most political science approaches recognise three main schools of thought, namely, Pluralist, Elitist (or Managerialist) and Marxist (class-based) interpretations (see, for instance, Alford and Friedland, 1985; Hay, 2006; Cudworth *et al.*, 2007). While Pluralism emphasises the centrality of social groups and political parties interacting and being represented by the state, Elitism underscores the asymmetries of power in society and the critical influence of political elites on state action. It is no secret that pluralist authors have been the main force behind mainstream political science, not necessarily in a coherent way but through the articulation of a variety of epistemologies and methods. Despite this, pluralists tend to agree that groups are more important for politics than individuals, and those groups complement, or are alternatives to, the state as mechanisms for production and distribution. Pluralism sees a separation between economic and political powers, which are core elements of liberal thinking and liberal societies. Pluralism continues to have a significant impact on the design of government policies and interventions, for instance, in the search for better forms of 'governance' and especially environmental governance, as we will see in Chapters 4 and 5. By contrast, elitist scholars dispute the basis of Western liberal assumptions about state, politics and civil society by arguing that society is primarily determined by the nature of its elite. The political elite are considerably detached from other social groups, which tend to adopt more passive behaviour. While classical elitism had a more rigid interpretation of the rise and decline of elites, modern elitist perspectives offer a more dynamic understanding of political competition and the organisation of the governing elite.

Notwithstanding the points made by pluralist and elitist authors, because the interventions of the state in the environmental arena are an integral feature of the expansion of capitalism and of the intensification of socioecological contradictions, a political ecology of the state can particularly benefit from a Marxist perspective on economy, society and nature. Marx was actually one of the first to study the sociology of the state (Badie and Birnbaum, 1983), in which he was followed by many other thinkers. The Marxist contribution is relevant to comprehend that capitalism is ultimately a socioeconomic system based on the double, interrelated exploitation of labour power and of the rest of socionature. Even though the association between capitalism and the state varies from country to country, the separation of the labourer from the means of production mirrors the alienation of individuals from their socionatural condition. In contrast to previous socioeconomic formations, such as feudalist countries or tribal nations, the capitalist management of socionature began to entail more than the procurement of raw materials and territorial power, as changes in the physical and symbolic configuration of nature were brought to the centre of capital accumulation. In Britain the strength of capitalist relations was historically connected with a relatively weaker and decentralised state (Wood, 2002), although it does mean that the British state has not been directly involved in the creation and resolution of environmental issues.

Considering that the reproduction of capitalism is a main sphere of state responsibility (Fine and Harris, 1979), a key dilemma of the contemporary state is how to exercise leadership on behalf of the wider society and simultaneously defend the interests of the stronger politico-economic groups. Contemporary environmental problems are not only derived from the failures of the state to arbitrate in the case of contradictory demands, but are also the result of the convergence of hegemonic interests in the organisation and running of the state. Mounting antagonisms within and between countries mean that the political equilibrium within which the state operates is always transitory and the action of the state reflects the fight for political hegemony. It is important to realise that Marxism also helps us to understand how relations and processes materialise in real life and how they are negotiated on an everyday basis, given that individuals not only suffer, but are also able to react, reinvent and often benefit from, at least to some extent, mainstream environmental policies. Consequently, a class-based perspective can provide a systemic and integrated investigation into the origins of environmental problems, whilst at the same time addressing structural differences, environmental injustices and social exclusion happening due to politicised socionatural interactions. The recognition of the fundamental importance of class inequalities does not obfuscate the existence of other forms of inequality associated with gender, race, age and religion, in the same way that justice goes beyond the simple redistribution of opportunities and also depends on political recognition and the ability to meaningfully intervene in public matters.

Although Marx only left a fragmented theory of the state, he understood it as the expression of social balance of power (but with a degree of relative autonomy), and elaborated on the effect of state forms and activities on the production and realisation of value (Ollman, 2003). As Marx (1913) unambiguously demonstrated, dominant class politics is neither necessary nor sufficient to ensure state reproduction of class relations, but the connections between state and class politics (including sub-sections of classes) are both contingent and non-linear. Economic issues are certainly prominent and trigger state action in multiple and unexpected ways, or, as put by Poggi (1978), under capitalism the economy subordinates and reduces the importance of other factors. Marx's attack on the bourgeois state was a humanist, radically democratic plan to release civil society from the inherent contradiction of an economy that keeps evolving through the strategic connections between state, hegemonic classes and class fragments. Following that initial Marxian elaboration, there has been a long debate about the connection between the economic and the political realms of state interventions, that is, the action of the state not only in relation to the interests of the capitalist class, but also in terms of the need to secure the cohesion of society as a whole (Clarke, 1991). The 'state question' has occupied a central position in Marxist discussions, which started with Engels (2010) and his contention that the state was a necessary evil for the exploitation of labour, and was followed by Lenin's (1932) claim that the bourgeois state must be smashed and replaced by a proletarian government and Gramsci's (1971) theory of the state based on force and consent.

The crisis of the welfare regime and the political unrest in the late 1960s provided an opportunity for Marxist authors to once again scrutinise the state from a class-based perspective (Jessop, 1977). More literal readings of Marx's and Engel's texts, such as the one by Mandel (1969), tried to situate the state as the product of class conflicts which operates on behalf of the ruling class. Other interpretations, such as the 'capital logic' school, attempted to derive the general form of the state from the capitalist mode of production (Holloway and Picciotto, 1978) and asserted that the state is the 'ideal collective capitalist' that operates independently of the actions of individual capitalists (Altvater, 1973). Because of the perceived ahistoricism of those claims, the 'regulationist' group tried to portray the state as the manager that compensates for the crisis between production and consumption (Aglietta, 1979), whereas Offe (1984) argued that the main problem with the welfare state was the contradiction between private ownership of industry and socialisation of production. The most notorious exchange of ideas in this period was between Miliband (1969), who claimed that the institutions comprising the state (such as parliament, government, policy and the judiciary) are colonised and primarily represent the capitalist class (that is, the state has significant power because it is a theoretical abstraction on behalf of the bourgeoisie), and Poulantzas (1967), who disagreed by saying that capitalist power permeates the state in a more systematic manner and that is what connects the structures of society and the state.

This disagreement appeared in many publications by the two authors, but the Poulantzian side of the debate seemed to provide a more dynamic theoretical model that accounts for the relative autonomy of the state apparatus from society and its role as a biased and contested arbiter. Jessop (1982) even ventured to find a middle ground (although in reality closer to Poulantzas) with his strategic-relational approach that depicts the state as an institutional ensemble of power centres that does not exist in isolation to wider disputes, but whose authority conforms with political economy and is embedded in social relations (note that Cudworth *et al.*, 2007, accuse Jessop of not being able to overcome the dualism between structure and agency or, what is worse, of deviating too far from the dialectical approach taken by Marx himself). A more general criticism put forward by the authors involved in this debate is that state theory is still a major 'hole' in the Marxist literature, either because of the tendency to either insist on the reification of the state (Abrams, 1988) or because of the propensity to abandon historico-materialist readings of the state and fall back into liberal or conservative legal theories (Macnair, 2006). Therefore, the key task for Marxist authors is to do away with the structuralist–functionalist boundaries of the Miliband–Poulantzas dispute (Clarke, 1991) and produce a more compelling theory of the tacit separation of the state from the capitalist class and its dealings with the other class-based demands, as well as to provide a better understanding of the constraints of state interventions and of the state's striving to foster unity in a fragmented society (including conflicts over socioecological issues). In this context, there is a clear opportunity to critically examine the rationality, function and justification of state action and to analyse how the production of environmental statehood has affected state discourse, configuration and legitimacy.

The main contribution of a Marxist approach should be the differentiation between appearance and reality, which is required to understand the fetishized, mystified elements of class struggle that permeates the state (Wells, 1981). The state can then be appreciated in its permanent, often contradictory, interplay with society and the rest of socionature, and any examination of the state needs therefore to encompass the broader, highly politicised evolution of the socionatural world. Three main points immediately emerge for further investigation, which will be dealt with in the following chapters, namely, the organisation and configuration of the state, the motivations and rationality of environmental responses, and the possibilities and limitations of state interventions. In relation to the first point, the formation of a dedicated nucleus of environmental policy and regulation within the state apparatus, especially since the 1970s, was both a reaction to pressures from civil society and the complaints of groups more directly impacted by environmental degradation, and also a response to inter-capitalist tensions and rising production costs due to ecological disruption and artificially created scarcities (e.g. shortage of land, water and biodiversity products). Even with their growing complexity, the environmental branches of the national state are normally instigated by foreign, multilateral organisations (UN, EU or the World Bank) and are actually devoid of any real power to tackle other larger politico-economic pressures (for example, the failure of public transport policies vis-à-vis the excessive use of the private car). The contradictions between various state policies mean that, in practice, the environmental agencies of the state – by necessity concerned with long-term, collective problems – are typically relegated to only a secondary level of importance.

Second, the main motivation of the state when intervening in environmental matters is to prevent socioecological disputes from escalating into class-based conflicts. Regardless of the lip service to environmental conservation adopted by many governments, the main commitments of the state are to the preservation of private property and production relations. Environmental policies and legislation serve, first and foremost, to systematise the access to, and ownership of, parts of socionature that have economic or political relevance, as well as to reduce production costs and uncertainties. Environmental responses by the state are also required because of the non-coincidence in time and space of the causes of environmental degradation and the actual impacts of those causes (i.e. a time-lag and space-lag phenomenon). It means that environmental protection is only an incidental objective and the primary reason for state action in environmental matters is to disguise the anti-commons requirement of capitalist production, which is conveyed through the management and circumstantial preservation of the same commons. The sustained action of the state, predominantly to serve the anti-commons demands of capitalist society, is the actual and genuine 'tragedy of the commons'. Coincidentally, it is interesting to observe that, although environmental regulation is primarily concerned with the amelioration of existing capitalist conditions, the specific interventions of the state can sometimes temporarily contradict that goal. Marx (1981) aptly identifies moments when interactions within the capitalist class lead to the dissociation between capital ownership and profit generation (i.e. the expansion of finance capital), which can in turn cause monopoly problems and require state intervention. The management of environmental issues has similarities to the management of economic imbalances, in the sense that it is also a situation where capitalism's drive for profit has to be contained, but as a measure to re-establish more general accumulation opportunities. A situation that involves the circumstantial "abolition of the capitalist mode of production within the capitalist mode of production itself" is a "point of transition" to new forms of production (Marx, 1981: 569).

Third, environmental statehood is permanently being reformulated and frequently undermined because of the intrinsic limitations of the state and the specific characteristics of environmental matters (which are unevenly distributed in space, unpredictably connected in time and scale, and associated with high levels of urgency and uncertainty). Many of the failures of public policies and environmental regulation originate not only in the political commitments of the state, but also in its incompetence when coordinating its own interventions. (Incidentally, this observation about the widespread incompetence of the state apparatus may be the answer to many of the exaggerated attacks made on the Marxist theory of the state by van den Berg, 2003.) The complexity of environmental issues has represented major challenges to the environmental ability of the traditional Westphalian state. In particular, given the global basis of environmental issues, a state may be responding not only to the hegemonic pressures of the national capitalist class, but also to fractions of classes in foreign countries and opposition groups in those same countries. As pointed out by Poulantzas (1967: 65), the Marxist conception of politics

> enables us to explain the possibility of *social formations in which there are 'disjunctions' between the class whose mode of production ultimately*

imposes its dominant political role on the one hand and the objective struc-
tures of the state on the other.... This will lead us to the following conclusion: there may be sizeable disjunctions between the politically dominant
class and the objective structures of the State.

The above three points allow us to start unveiling the environmental dimension of the contemporary state, in particular the specific features of its present-
day responses to environmental problems formulated under the aegis of
'governance' (rather than simply 'government'). Governance, instead of conventional government, is described as the pursuit of more flexible strategies and
mechanisms of public administration in order to accomplish policy goals, realise
values and manage environmental risks and impacts. It includes a range of regulatory processes, incentives and institutional changes aimed at raising awareness,
influencing personal and group behaviour and involving social actors in decision-
making. Environmental governance is supposedly different to conventional
environmental management (i.e. operational activities for meeting specific
targets) in the sense that it has a combined focus on regulatory institutions (laws,
policies), informal institutions (power relations and practices) and organisational
structures and their efficiency. The aims of the governance agenda are expected
to be achieved through the re-regulation of conservation and of the use of natural
resources, which amalgamates state-oriented and market-oriented approaches.
The most common interpretation of governance in international literature relies
on the Foucauldian notion of governmentality and its focus on bio-power of the
state as demonstrated by its plurality of interventions in health, hygiene and
environmental issues. Power is seen as being dispersed throughout society and is
a positive force in the creation and change of behaviour (Foucault, 1977), which
in the environmental arena leads to the formation of new expert knowledge,
social practices, regulatory approaches and subjectivities (Agrawal, 2005).
However, the analysis informed by Foucault's governmentality seems to offer
little assistance to understanding state action and failures, given that it essentially diverts our attention away from the state. A more accurate assessment of
the peculiarities of environmental governance should, therefore, focus on the
sociopolitical relations that permeate, and are encouraged by, the state, instead
of restricting the analysis to explanations that tend to minimise the socioecological centrality of the state as both a mediator and a champion of the demands of
groups, classes and class-fractions. As already mentioned above, the disagreement between Hegel and Marx, where the former strived to perfect the emerging
state of industrial capitalism (even before the bourgeoisie became hegemonic in
Prussia) and the latter called for the 'dismantlement' of the (capitalist) state, has
huge, although somewhat surprising, relevance nowadays to the examination of
environmental governance and other connected approaches. The tortuous evolution to more flexible environmental statehood has reproduced, on the small scale
and at the sectoral level, the clashes between those two main thinkers of the
nascent urban–industrial capitalist world, mainly because the grave environmental contradictions of modern-day capitalism have close parallels with the

socioeconomic and political challenges of the early nineteenth century. In other words, these are two major crises of the same socioeconomic system, only manifested in different ways across different spatial and temporal circumstances. A close scrutiny of environmental governance, for instance, reveals a disturbing element of political Hegelianism, specifically in terms of the constant search for an elevated rationality over group interests and the reinforcement of the legitimacy of state power. The main goal of the contemporary governance agenda is the transition from rigid, monothematic environmental regulation into more responsive interventions, which betray the influence of Hegelian political thinking.

Marx's critique of Hegel has, therefore, major significance for the examination of the environmental agenda of the contemporary (capitalist) state, particularly in challenging context of the early twenty-first century, something that will be further analysed in detail in Chapter 6. In contrast to the idealism of Hegel's political model, Marx rejected the view that the state could be described as an all-encompassing political community functioning according to an ethical appeal and acting as the fulfilment of reason. On the contrary, the state operates within the conflict between the interests of the individual and those of the community, but always takes sides in favour of the stronger classes (to the extent that "struggles within the State ... are merely the illusory forms in which the real struggles of the different classes are fought out among one another", Marx and Engels, 1974: 53). In the well-known preface to *A Contribution to the Critique of Political Economy*, Marx (1975) shows how Hegel, through a skilful handling of ethics and dialectics, ultimately reduced civil society to economic society. This resembles the reduction of the complexity of ecosystems to the narrow language of money and the techno-bureaucracy of computer models that characterise today's prevailing approaches to environmental regulation and management. Hegelianism has largely inspired the neopragmatism of ecological modernisation (Davidson, 2012), whose "normalisation of practices tends to obscure their philosophical premises and the separation of theory from practise, finance from politics, policy from implementation" (Irwin, 2007: 648).

Marx fundamentally objected to the Hegelian claim that the state should be the idealised domain of reason, concord and actualised freedom. For Marx, capitalism becomes the human basis of a state that uses mystification, through religion and politics, to maintain the basis of socioecological exploitation. In that regard, Hegelian political theory provided the necessary ideological and managerial tools to allow the dissimulated expansion of capitalist socioeconomy over socioecological systems, as is the case with the current agenda of environmental governance. The political platform of Hegel was centred on the spreading of universalisms among particularities and a subtle legitimacy of prevailing socioeconomic relations through the homogenisation of politics and the promotion of reason. Needless to say, ideological disputes, together with material processes, help to shape human interventions and state action and, in the end, represent important 'geography makers' (Mann, 2009). Moreover, the continuities and divergences between Marx and Hegel are complex and have been hotly debated for more than a century. Of course, most of Marx's analysis of Hegelian political thinking was very selective and done early in his life, and there was never an opportunity to consolidate it in the broader context of his study of economy and society. The young Marx was a member of the group of Left-Hegelians and was influenced by the humanism of some of its key members, such as Bauer and Feuerbach. The group debated on Prussian politics, religion and sociology, criticised the idealisation of the state by the old Hegelians, and attacked Hegel himself for failing to deal with what they called 'the living man' (i.e. the sensuous person who actually thinks, feels and creates his or her own existence over any generalisation).

The young Hegelians strongly disagreed that the state could claim its legitimacy on religious grounds, because the corruption and despotism of the state was actually considered the embodiment of religion. Whilst involved in these discussions, Marx increasingly radicalised his political position, rejecting Hegelian idealism and Left-Hegelianism, and shifted his criticism away from religion and towards private property and the ownership of capital (Berlin, 1978). Hegel leaves behind the notion of social contract of Rousseau and Kant in favour of a constitutional formula based on the rational state, as the spiritualised form of the 'Idea' (Chaskiel, 2005), but Marx (1970) rebuffed exactly such argument that reality was a predicate of the 'Idea' (i.e. the Idea as the actual subject of historical development). Marx (1975) also argued that the contribution of Feuerbach lay in reasserting the importance of the material world (in contrast to the neo-Hegelian idealism that prevailed in German thought at that time), but that Feuerbach's vision of nature was essentially ahistorical and, ultimately, reactionary. The influence of Hegel on Marx has been a matter of fierce scholarly disagreement and has given rise to many interpretations. For instance, McCarney (1999) sustains that the relationship between Hegel's and Marx's political systems was one of misunderstandings and misappropriations, unnoticed slippages, wrong turnings and lost opportunities, while for Levine (2009), Marx did not perceive the materialist aspects of Hegel's presentation of civil society and exaggerated the separation between state and civil society in the Hegelian construct. Regardless of the correctness of those criticisms, it is clear that even if Marx always remained a Hegelian (at least in terms of his analytical and interpretative methods), he tried to bring politics to the centre of social and economic analyses and set his assessment of the state on the course of transformative revolution and comprehensive social inclusion. More importantly here, Marx's struggles to overcome the Hegelian mystification of the state, and his effort to move beyond Hegel's immanence–transcendence approach, have close parallels with the removal of the alienated basis of environmental statehood.

Starting from the socioecological aspirations, possibilities and limitations of the contemporary state, the next few chapters will present a politico-ecological assessment – informed by Marx's analysis of capitalism – that not only captures the state as the main locus of environmental ideology and transformative capacity, but also situates the ecological properties of the state within the wider interactions between social classes, economic policies and grassroots demands. The contemporary (capitalist) state will be examined according to its crucial, unavoidable ontological dilemma: it can neither dissociate itself from class politics nor prevent class conflicts from affecting the long-term viability of socioecological systems. The reflexivity of the state's socioecological involvements needs to be part of a wider critique of class conflicts and politico-economic asymmetries that strongly influence socionatural interactions. As affirmed by Scholsberg (2007: 199), there is still "very little reflexivity of the state" in relation to its role as one of the main causes of the formation of unjust environmental situations. The study of environmental politics has so far focused too much on the outcome of public policies and on the conflicting associations between groups, but too little has been said about the intricacies of the state apparatus as a key force behind the production of nature or, what is even more important, on how nature has meant a great deal for the reorganisation of the state as a result of the overall pressures to stabilise the political and economic contradictions of capitalist relations of production and reproduction.

The chapters in this book are organised in a sequence that provides an analysis of the different models of environmental statehood adopted in specific historical moments. The political ecology of the state is located in the multiple contradictions of environmental statecraft, that is to say, on how environmental statehood and the practice of environmental regulation have evolved and in an attempt to reduce socioecological problems within the narrow framework of private property and socionatural exploitation. In addition, the limits to the responses to environmental problems are also manifestations of a deeply ingrained legacy of dealing with public matters. The examination of these problems will benefit from the assessment of concrete and interrelated examples from northern and southern countries. Unlike the claims made by Whitehead *et al.* (2007), the main question for political ecology is not only the relationship between national states and global commons, but between the multiple spheres of the state apparatus that tries to control and regulate the 'more-than-human' dimensions of the world. It is not just the national state and the global commons that matter, but both the state and the commons consist of connections across different scales, localities and times (i.e. the responsibility for environmental protection and for resolving associated conflict extends from the national state structure to local authorities and multilateral international agencies).

The current book aims to enrich poststructuralist and materialist analytical positions by specifying ways in which power is enacted and by analysing political struggles related to the access, use and conservation of socionature. This opening chapter has already offered an introductory discussion of the direction environmental statehood is heading in and of the socioecological repercussions of state interventions. The state was situated at the centre of complex socioecological relations, as the main entity responsible for the transformation (i.e. production) of nature and as the biased mediator of politicised environmental issues. It has been argued that environmental policy and regulation has been put in place by the contemporary (capitalist) state according to a utilitarian logic, which has typically been reactive and dominated by wider politico-economic priorities.

This control of environmental agendas by politico-economic interests gives strong justification for a Marxist interpretation of environmental statehood. Even if one of Marx's unfulfilled ambitions was to write more directly about the connection between state and capitalism, in his existing body of work we can find a clear understanding of statecraft and statehood as the expression of social power, particularly the power of the dominant class, as well as the effect of state forms and activities on the production and realisation of value. These observations obviously need to be seen in tandem with Marx's elaboration on society's interdependencies with nature and his advocacy of the withering away of the (capitalist) state (included in the latter part of the book).

Chapter 2 elaborates on some fundamental concepts required for an eco-Marxist interpretation of the achievements and constraints of environmental statehood. It will initially expand on the definition of environmental statehood and the narrow rationality of state interventions (within the broader irrationality and contradictions of socioeconomic relations). The prolongation of an irrational environmental rationality has close parallels with the core elements of Western modernity and its dichotomic separation of society from the rest of socionature (Engels, 1954). The modern world was constructed around the Western type of individualism, which notoriously detaches human beings from their socioecological condition and places them above the rest of nature, as well as increasingly immersing them (i.e. human beings) in self-centred relations of production and consumption. Environmental statehood happens to be a contingent institutional arrangement adopted in a specific historico-geographical context under the influence of the ideological basis of capitalism and modernity. Ultimately, any "distribution whatever of the means of consumption is only a consequence of the distribution of the condition of production themselves" (Marx, 2001: 21). During the last century, three different environmental statehoods were put in place, encapsulating three characteristic interpretations and reactions to environmental problems (namely: early, conventional and flexible forms of environmental statehood), which will be analysed in the following chapters. The operationalisation of environmental statehood happens through the formation of transitory state-fixes, which contain the agencies, norms and discourses articulated according to concrete political and socioecological circumstances. A state-fix aims to respond to emerging environmental situations that affect society and the economy, but only within the limits of the hegemonic political and economic priorities that shape environmental statehood. In order to comprehend the specificities of each state-fix, Chapter 2 will put forward the idea that the interconnections between nature, state and society require a proper conceptual and empirical treatment. It is then argued that a trialectical interpretation would help us to appreciate the uniqueness of environmental statehood as a moment, a particular compromise that is part of, but also helps to shape, an inherently socionatural reality.

Chapter 3 deals with the organisation of the early and conventional models of environmental statehood, which, with their corresponding state-fixes, were responsible for dealing with natural resources and socioecological issues during most of the twentieth century. Environmental statehood evolved from the rationality behind early legislation at the turn of the century to a broader and more intricate package of environmental laws and dedicated agencies established in the post-Second World War period. The chapter argues that environmental statehood progressed in a way that largely replicated the organisation of the capitalist state more than a century earlier. The political philosophy of some key European authors has either served as inspiration for the evolution of environmental statehood, or their political thinking can be considered a metaphor of the actual processes of institutional change. If nothing else, referring to the different schools of philosophical thought provides a relevant analogy that helps us to understand the achievements and limitations of the various phases of environmental statehood. This chapter will also demonstrate that the main pillars of conventional environmental statehood can be found in the ideas of Hobbes, especially his elaboration on authority and containment of natural brutality. Regulatory approaches began to intensify in the 1960s and 1970s, when new dedicated agencies and novel legislation were informed by Hobbesian thinking. The problems of such centralised forms of environmental statehood became increasingly evident and as a result, since the 1980s, there were calls for regulatory flexibility. In this process, there was a (partial) transformation of the control of environmental degradation into the profitability of environmental degradation. The discussion is followed by a case study in Rio de Janeiro, which shows how the employment of procedures for the allocation of resources, communication of information and management of risks using the conventional forms of environmental statehood were inadequate and meant further conflict.

Chapter 4 will discuss the conceptual and operational details of the transition to a more flexible and responsive type of environmental statehood. With the perceived failures of the environmental agenda espoused by the welfare or welfare-developmentalist state,[5] a new model of environmental regulation was introduced through concepts such as sustainability, public participation, adaptive management and governance (rather than government). This process has had contingent connections with the expansion of neoliberal policies and the neoliberalised reform of the state apparatus. Conventional environmental statehood was replaced by market-friendly and market-dependent strategies that expanded the capitalist reasoning to domains previously beyond the reach of the market and private property relations. This transition benefited from Kant's calls for higher levels of reason and the advance of political rights. The rearrangement of environmental statehood along Kantian lines not only preserved the fundamental anti-commons trajectory of the capitalist state, but also provided new opportunities for the circulation and accumulation of capital directly from environmental conservation. These points are illustrated with an analysis of the transition to a more market-friendly management of water in Brazil after the introduction of a new regulatory framework in the 1990s.

Chapter 5 expands on one particular element of the transition to a more flexible environmental statehood, namely the progressive dominance of neoliberal institutions over socioecological systems. This emerged as both a response to the ecological crisis of Keynesian capitalism and as novel business opportunities.

The late Neil Smith provocatively argued, paraphrasing Habermas, that "neoliberalism is dominant but dead".[6] This claim serves to illustrate the paradoxical situation of neoliberalism in the early twenty-first century, in the sense that it continues to have a determining influence on policies and socioeconomic relations, but also shows increasing signs of failure. Neoliberal policies entail the destruction and reconstruction of previous forms of environmental statehood, but in practice there were some activities which the state was never able to delegate to the market, as the economic elite would have preferred. The controversies of the neoliberalisation of nature that underpin the search for more flexible environmental statehood are demonstrated with examples from the implementation of the Water Framework Directive in the European Union and also the reform of water services since 1990 in Lima, Peru. In both cases, the state was the main promoter and advocate of nature neoliberalisation and had to constantly mediate in old and new conflicts created by neoliberalising policies.

The case studies included in Chapters 3, 4 and 5 are the result of fieldwork conducted between 2008 and 2011 in various countries. The intricate connections between society, state and the rest of nature inevitably posed serious problems for research on the political ecology of the state in some of these countries. This called for a comprehensive analytical approach that was able to capture 'the whole' – the totality of relations (Ollman, 2003). Marx claimed that it was necessary to start with the 'real concrete' (the world in its intricacy) and proceed through progressive 'abstractions' (that is, breaking reality into intelligible units) to the 'thought concrete' (the understood whole, critically interpreted). As Marx (1973: 101) put it, the "concrete is concrete because it is the concentration of many determinations, hence the unity of the diverse". In an attempt to address the totality of the concrete, our investigation of the political ecology of the state consisted of 'embedded case studies' with more than one sub-unit of analysis to understand a complex whole. The warning message of Thompson (1966: 204), whose "objection to the reigning academic orthodoxy is not to empirical studies per se, but to the fragmentation of our comprehension of the full historical process", was certainly taken into account in the design of the case studies in order to avoid the same mistakes. Embedded case studies are particularly useful in political geographical investigation because they can be employed to investigate large processes that involve many individual sectors and organisations. Our case studies started with the consideration of sub-units of social action and then scaled them up to identify common patterns over larger geographical areas. The research explored interests and behavioural patterns in various geographical locations and stakeholder sectors, as well as the institutional framework in which they operate.

Initially, the research effort included scoping interviews with key informants and academics involved in environmental regulation and urban policy-making. Based on these preliminary interviews and secondary information, databases of public and non-governmental sectors were developed to guide further interviews, the analysis of documentation and additional collection of background information. By mapping the various organisations, their discourse and stated aims, it was possible to compare intra- and inter-group differences and a range of alliances or disputes. A long list of semi-structured interviews was then carried out and provided a broad range of opinions in favour of, or against the process of, institutional reforms, including the opinions of local residents, regulators, policy-makers and parliamentarians, NGO activists, workers in public companies and representatives of multilateral agencies. Interview respondents were chosen from an array of organisations that represented multiple interests in the environmental sector. Additional data were gathered from close engagement with local residents (i.e. participant observation), which was instrumental in capturing their beliefs, practices and subject positions. Public events sponsored by both governmental and non-governmental entities were also attended during the fieldwork periods. Special effort was made to participate in as many different types of activities as possible in order to build a solid understanding of public debate and conflicts of interest. Official documents were also consulted at public and private libraries, as well as reports provided by the various organisations visited during the research campaigns. Finally, keeping up with media coverage of socioecological issues and the daily consultation of the main web-based newspapers extended beyond the period in the field.

Building on the conceptual and empirical evidence included in the preceding parts of the book, Chapter 6 will put forward the main thesis, that is, that Hegelian thinking is an unexpected influence on the implementation of neoliberal and, increasingly, post-neoliberal mechanisms of environmental regulation by the capitalist state. The Hegelian notion of the state as the ultimate promoter of collective reason and political ethics has permitted the expansion of state responsibilities to the realm of socionatural interactions, going far beyond the simple appropriation of natural resources. Hegel's political theory goes beyond Hobbesian (coercive authority) and Kantian (rational freedom) concepts and becomes an endorsement of more malleable and legitimated state interventions. It will be shown that Hegel constantly makes reference to a cautious, safe adaptation and to the need to adjust to new circumstances within strict boundaries of rationality, legitimacy and hierarchy. The Hegelian conceptualisation of the modern state – idealised as the immanence of state power and reason, and the essential role of political and economic differences as historical forces – provided the necessary justification for the renewal of environmental statehood. In the more recent organisation of the state, the role of the Hegelian crown was replaced by other public figures (prime ministers and presidents), the estates became large parliaments and the executive is now a vast structure with manifold branches and agencies.

The mystification of the economic and sociopolitical objectives of contemporary environmental regulation constitutes the actualisation of Hegel's plan for the modern state. In the end, this becomes a clever and highly useful dissimulation of political and socioeconomic alienation through the idealisation of both 'people' and the 'state'. If Hegel anticipated the environmental regulation put forward at the end of the twentieth century, he also created a lasting trap that largely limits the possibility of establishing a truly democratic socionatural relationship. In providing novel justification for a system that is inherently self-destructive – based on the double exploitation of society and the rest of nature – the soft-violence of the Hegelian state turns into the maximisation of violence. Even the style and prose of Hegel, fraught with contradictory claims and obscure statements, are quite appropriate for the relativism, post-productivism and mass waste of capitalist societies. As Marx prudently observed, the Hegelian state model presupposes an individual that is separated from his or her social bonds, as well as the perpetuation of the sources of alienation and exploitation. This comment is as valid now as it was in the 1840s when Marx was striving to do away with the powerful influence of Hegelian political theories.

Finally, after the discussion on the political ecology of the state and the evolution of environmental statehood, the last chapter will consolidate the lessons learned and recommend the need to search for deeper transformations and radically different state praxis. In Chapter 7, it will be seen that some of the main functions of the state continue to be the stabilisation of capitalism's self-destructive disposition, the invigoration of accumulation mechanisms and the guarantee of socionatural exploitation opportunities. It will be argued that the centrality of private property still represents one of the chief reasons for the existence of the contemporary state, but private property institutions also created a range of unsolvable contradictions to do with the interconnections between humans and the rest of socionature. Due to the contradictory basis of the capitalist state (i.e. it is both promoter and moderator of capitalist relations of production and reproduction), its responses are only partially effective and, what is more, always reproduce an asymmetric distribution of gains and losses that penalises vulnerable social groups and socioecological systems and ultimately threatens the long-term viability of capitalism. The mainstream forms of environmental statehood, one of the most dynamic and controversial responsibilities of the contemporary state, have been fraught with ambiguities and are evidently unable to face up to local and global socioecological contradictions. The more the capitalist state deals with environmental issues, the more it sustains the exploitation of nature and widens the gap between society and its ecological condition, just as it does between different groups and classes. The little that has been achieved in terms of environmental conservation and impact mitigation is *in spite of* the state rather than *because of* the state.

Notes

1 For the purpose of this book, the expressions 'environmental' and 'socioecological' are used interchangeably to describe the interconnections, interdependencies and co-evolution between society and the rest of socionature.
2 The contemporary (capitalist) state is more than just the national state, but branches into municipal and provincial authorities, as well as international, regional and multilateral organisations.
3 This hegemonic anti-commons propensity of the state has, however, left some space for the remnants of commons, such as the communal meadows of Europe, international seawaters and collectively managed rainforests. These commons persist, and sometimes

flourish, even in the globalised market economy of the early twenty-first century. Instead of exotic institutions forgotten or tolerated within the hegemonic spread of private property relations, the remaining commons play an important role for the sustainability of wider capitalist relations, as sources of biodiversity, ecosystem services and survival opportunities to destitute social groups (incidentally preventing them from attacking private property).

4 Institutions, which can be defined as systems of prevalent social rules that structure social interactions (Hodgson, 2006), are complex phenomena, whose reproduction is incomplete, provisional and unstable, and which co-evolve with a range of other complex emergent phenomena (Jessop, 2001). Like all social institutions, environmental institutions – such as property rights over resources and conservation approaches – are subjective, path dependent, hierarchical and nested both structurally and spatially. Environmental institutions are embedded within the cultural, social, economic and political context. This means that the institutional reforms associated with environmental management cannot be seen in isolation, but are part of the larger agendas of state reconfiguration and socioecological disputes. The reform of institutional arrangements has a marked political dimension, when we consider that hegemonic values within society reflect prevailing power structures that are legitimised in and by institutions (Cumbers *et al.*, 2003).

5 The welfare (in northern, industrialised countries) or welfare-developmentalist (in southern, post-colonial countries) state was the main state formation during most of the twentieth century. It was (heavily) influenced by Keynesian–Fordist economics, which were adapted to each specific national circumstance (Ioris and Ioris, 2013).

6 This reference was made at the annual meeting of the Association of American Geographers held in 2010 in Washington, DC. It is also a humble homage to Prof. Neil Smith (who passed away in 2012), with whom the author had the opportunity to learn about, and discuss, political ecology during our time together at the University of Aberdeen.

Bibliography

Abrams, P. 1988. Notes on the Difficulty of Studying the State. *Journal of Historical Sociology*, 1(1), 58–89.

Aglietta, M. 1979. *A Theory of Capitalist Regulation: The US Approach*. Verso: London.

Agrawal, A. 2005. *Environmentality: Technologies of Government and the Making of Subjects*. Duke University Press: Durham and London.

Alford, R.R. and Friedland, R. 1985. *Powers of Theory: Capitalism, the State, and Democracy*. Cambridge University Press: Cambridge.

Altvater, E. 1973. Some Problems of State Interventionism. *Kapitalistate* 1, 96–108 and 2, 76–83.

Arrighi, G. 1994. *The Long Twentieth Century: Money, Power, and the Origins of Our Times*. Verso: London and New York.

Badie, B. and Birnbaum, P. 1983. *The Sociology of the State*. Trans. A. Goldhammer. University of Chicago Press: Chicago and London.

Berlin, I. 1978. *Karl Marx*. 4th edition. Oxford University Press: Oxford.

Bernstein, S. 2001. *The Compromise of Liberal Environmentalism*. Columbia University Press: New York.

Callinicos, A. 2007. Does Capitalism Need the State System? *Cambridge Review of International Affairs*, 20(4), 533–49.

Chaskiel, P. 2005. De Rousseau à Marx: Les Métamorphoses du Peuple. *Hermès*, 42, 32–7.

Clarke, S. (ed.). 1991. *The State Debate.* Macmillan: Basingstoke and London.

Conca, K. 1993. Environmental Change and the Deep Structure of World Politics. In: *The State and Social Power in Global Environmental Politics*, Lipschutz, R.D. and Conca, K. (eds). New York: Columbia University Press, pp. 306–26.

Conca, K. 2006. *Governing Water: Contentions Transnational Politics and Global Institution Building.* MIT Press: Cambridge, Mass. and London.

Cudworth, E., Hall, T. and McGovern, J. 2007. *The Modern State: Theories and Ideologies.* Edinburgh University Press: Edinburgh.

Cumbers, A., MacKinnon, D. and McMaster, R. 2003. Institutions, Power and Space: Assessing the Limits to Institutionalism in Economic Geography. *European Urban and Regional Studies*, 10(4), 325–42.

Davidson, S. 2012. The Insuperable Imperative: A Critique of the Ecological Modernizing State. *Capitalism Nature Socialism*, 23(2), 31–50.

Dobson, A.N.H. 1998. *Justice and the Environment: Conceptions of Environmental Sustainability and Dimensions of Social Justice.* OUP: Oxford.

Dryzek, J.S. and Lester, J.P. 1989. Alternative Views of the Environmental Problematic. In: *Environmental Politics and Policy: Theories and Evidence*, Lester, J.P. (ed.). Durham and London: Duke University Press, pp. 314–30.

Eckersley, R. 1992. *Environmentalism and Political Theory: Toward an Ecocentric Approach.* SUNY Press: Albany, NY.

Eckersley, R. 2004. *The Green State: Rethinking Democracy and Sovereignty.* MIT Press: Cambridge, MA and London.

Engels, F. 1954. *Dialectics of Nature.* Trans. C. Dutt. Foreign Languages Publishing House: Moscow.

Engels, F. 2010. *The Origin of the Family, Private Property and the State.* Penguin: London.

Fine, B. and Harris, L. 1979. *Rereading Capital.* Macmillan: London and Basingstoke.

Foucault, M. 1977. *Discipline and Punish: The Birth of the Prison.* Trans. A. Sheridan. Allen Lane, Penguin: London.

Gramsci, A. 1971. *Selections from the Prison Notebooks.* Trans. Q. Hoare and G.N. Smith. Laurence and Wishart: London.

Habermas, J. 1987. *The Theory of Communicative Action: The Critique of Functionalist Reason.* Vol. 2. Trans. T. McCarthy. Polity Press: London.

Hay, C. 2006. (What's Marxist about) Marxist State Theory. In: *The State: Theories and Issues*, Hay, C., Lister, M. and Marsh, D. (eds). Palgrave Macmillan: Houndmills, pp. 59–78.

Hayward, T. 1998. *Political Theory and Ecological Values.* Polity Press: Cambridge.

Hodgson, G.M. 2006. What are Institutions? *Journal of Economic Issues*, 40(1), 1–25.

Holloway, J. and Picciotto, S. (eds) 1978. *State and Capital.* Edward Arnold: London.

Ioris, R.R. and Ioris, A.A.R. 2013. The Brazilian Developmentalist State in Historical Perspective: Revisiting the 1950s in Light of Today's Challenges. *Journal of Iberian and Latin American Research*, 19(1), 133–48.

Irwin, R. 2007. The Neoliberal State, Environmental Pragmatism, and Its Discontents. *Environmental Politics*, 16(4), 643–58.

Jessop, B. 1977. Recent Theories of the Capitalist State. *Cambridge Journal of Economics*, 1(4), 353–73.

Jessop, B. 1982. *The Capitalist State: Marxist Theories and Methods.* Blackwell: Oxford.

Jessop, B. 2001. Institutional Re(turns) and the Strategic-relational Approach. *Environment and Planning A*, 33, 1213–35.

Lefebvre, H. 2008. *Space, Difference, Everyday Life*, Goonewardena, K., Kipfer, S., Milgrom, R. and Schmid, C. (eds). Routledge: New York.

Lenin, V. 1932. *State and Revolution*. International Publishers: New York.

Levine, N. 2009. Hegelian Continuities in Marx. *Critique: Journal of Socialist Theory*, 37(3), 345–70.

Luxemburg, R. 1951 [1913]. *The Accumulation of Capital*. Trans. A. Schwarzschild. Routledge and Kegan Paul: London.

Macnair, M. 2006. Law and State as Holes in Marxist Theory. *Critique*, 34(3), 211–36.

Mandel, E. 1969. *Marxist Theory of the State*. Merit Publishers: New York.

Mann, G. 2009. Should Political Ecology be Marxist? A Case for Gramsci's Historical Materialism. *Geoforum*, 40(3), 335–44.

Marx, K. 1913 [1852]. *The Eighteenth Brumaire of Louis Bonaparte*. Trans. D. de Leon. Charles H. Kerr & Company: Chicago.

Marx, K. 1970 [1843]. *Critique of Hegel's 'Philosophy of Right'*. Trans. A. Jolin and J. O'Malley. Cambridge University Press: Cambridge.

Marx, K. 1973 [1857–8]. *Grundrisse*. Trans. M. Nicolaus. Penguin and New Left Review: London.

Marx, K. 1975. *Early Writings*. Trans. R. Livingstone and G. Benton. Penguin and New Left Review: London.

Marx, K. 1976 [1867]. *Capital: A Critique of Political Economy*. Vol. 1. Trans. B. Fowkes. Penguin: London.

Marx, K. 1981 [1894]. *Capital: A Critique of Political Economy*. Vol. 3. Trans. D. Fernbach. Penguin: London.

Marx, K. 2001 [1875]. *Critique of the Gotha Programme*. The Electric Book Company: London.

Marx, K. and Engels, F. 1974 [1845–6]. *The German Ideology*. Trans. C.J. Arthur. Lawrence and Wishart: London.

McCarney, J. 1999. Hegel's Legacy. *Res Publica*, 5(2), 117–38.

Miliband, R. 1969. *The State in Capitalist Society*. Basic Books: New York.

Mol, A.P.J. and Sonnenfeld, D.A. (eds). 2000. *Ecological Modernisation around the World: Perspectives and Critical Debates*. Frank Cass/Routledge: London and New York.

Offe, C. 1984. *Contradictions of the Welfare State*. Hutchinson: London.

Ollman, B. 2003. *Dance of the Dialectic: Steps in Marx's Method*. University of Illinois Press: Urbana and Chicago.

Paehlke, R. and Torgerson, D. (eds). 2005. *Managing Leviathan: Environmental Politics and the Administrative State*. 2nd edition. Broadview Press: Toronto.

Painter, J. 2006. Prosaic Geographies of Stateness. *Political Geography*, 25(7), 752–74.

Paterson, M., Doran, P. and Barry, J. 2006. Green Theory. In: *The State: Theories and Issues*, Hay, C., Lister, M. and Marsh, D. (eds). Palgrave Macmillan: New York, pp. 135–54.

Poggi, G. 1978. *The Development of the Modern State: A Sociological Introduction*. Stanford University Press: Stanford, CA.

Poulantzas, N. 1967. Marxist Political Theory in Great Britain. *New Left Review*, 43, 57–74.

Poulantzas, N. 1978. *State, Power, Socialism*. Trans. P. Camiller. Verso: London and New York.

Scholsberg, D. 2007. *Defining Environmental Justice: Theories, Movements, and Nature*. Oxford University Press: Oxford.

Smith, A. 2008 [1776]. *Wealth of Nations*. Oxford University Press: Oxford.

Thompson, E.P. 1966. *The Making of the English Working Class*. Vintage Books: New York.

van den Berg, A. 2003. *The Immanent Utopia: From Marxism on the State to the State of Marxism*. Transaction Publishers: New Brunswick and London.

Wells, D. 1981. *Marxism and the Modern State: An Analysis of Fetishism in Capitalist Society*. Harvester Press: Sussex and Humanities Press: New Jersey.

Whitehead, M., Jones, R. and Jones, M. 2007. *The Nature of the State: Excavating the Political Ecologies of the Modern State*. Oxford Geographical and Environmental Studies. OUP: Oxford.

Wood, E.M. 2002. *The Origin of Capitalism: A Longer View*. Verso: London and New York.

2 The reason and the purpose of environmental statehood

Environmental statehood as irrational rationalism

The multiple efforts undertaken by the contemporary state to respond to, and prevent, additional environmental problems were described in the previous chapter as the gradual assemblage of new, dedicated social institutions. Those institutions, together with environment-related discourses and strategies, formed what was already defined as *environmental statehood*. It involves the application of specific forms of socioecological interaction and ideological approaches to the control and regulation of ecosystems and territorial resources. Statehood must be interpreted as a dynamic process that is constantly reformulated according to homogenisation and particularisation pressures, and in accordance with the balance of political power (Brenner, 2004). Because of the growing relevance of environmental statehood in the politico-economic agenda since the early twentieth century, practically all countries in the world have allocated responsibilities to specific branches of the state in charge of overseeing environmental matters. Consequently, the evolution of environmental statehood is important not just in terms of addressing environmental issues per se, but it has become a key ingredient of the wider process of statecraft. The most noticeable change in the configuration of the state apparatus derived from the introduction of environmental statehood – which will be defined below as *state-fix* – comprises the work of policy-makers, regulatory agencies and law enforcement units, who employ a range of mechanisms including thresholds, charges and sanctions, as well as technical guidance, public education and economic incentives. Interestingly, in many cases, environmental statehood is not adopted because of pressures from national civil society or from the groups affected by environmental degradation, but due to the influence of international organisations and multilateral agencies (for example, United Nations divisions and development banks that have environmental risk assessment as part of their lending criteria). The recognition that environmental statehood is the result of both endogenous and exogenous demands reveals a great deal about the actual purpose and the political constraints that limit the resolution of local, national and transboundary questions.

Based on the above, the current chapter will expand on the basis of environmental statehood and will make reference to other related concepts needed for the examination of its failures and achievements. The first observation is that, considering its class-based commitments, there are major obstacles to the ability of the state to adequately respond to mounting environmental questions. As affirmed by Harvey (2006), the role of the capitalist state is quite complex, given that it must simultaneously centralise and decentralise responsibilities, as well as perform the conflicting duties of both a regulator and an entrepreneur. The state is required to act as a mediator of socionatural relations and the producer of new forms of interaction (according to specific historico-geographical circumstances and the range of pressures exerted on the state apparatus). Obviously, the format and implementation of environmental regulation is not a linear or predetermined process, but it unfolds according to the concreteness of hegemonic political pressures and the level of resistance from grassroots groups. Hence the importance of not only taking into account the diversity of state formations without reducing the analysis of environmental statehood merely to the external configuration, but also focusing on the essential characteristics of state action (Gerstenberger, 2011). On the one hand, environmental statehood is a demonstration of the creativity and resilience of existing socioeconomic structures, which have been able to contain, at least in the short term, many environmental threats and transform barriers into market opportunities. On the other hand, the evolution of environmental statehood is a fluid phenomenon that is articulated according to political asymmetries and socioeconomic inequalities.

It is not difficult to verify that the environmental agenda of the state – including the intervention of ministries, commissions and regulatory agencies, the approval and enforcement of legislation, policies and regulatory instruments, etc. – clearly bears the hallmark of power struggles directly and indirectly related to environmental issues. For example, the stronger social groups often form political blocks capable of rejecting the approval of more stringent environmental legislation demanded by critical NGOs and organised communities. In a society with antagonistic class-based relations, the political equilibrium within which environmental statehood operates is unstable and the actions of the state reflect pressures associated with the uneven opportunities available to different social groups. As argued by Engels (2010), the state was invented at the dawn of civilisation as an institution needed to secure the riches of some individuals against the previous communitarian traditions. In that context, the leadership of the capitalist state is not something given in advance or automatically conveyed to all involved parties, but any political settlement is itself formed through clashes, disputes and dialogue. The configuration and functioning of the state is "a relationship of forces, or more precisely the material condensation of such a relationship among classes and class fractions" (Poulantzas, 1978: 128), which expands through different scales, from local interactions to the realm of international relations (Agnew, 2001; Brand and Görg, 2008; Wissen, 2009).

One key dilemma for the contemporary (capitalist) state is precisely the need to advance environmental statehood in a way that seems to represent the whole society whilst simultaneously preserving the interests of the stronger politico-economic groups in charge of the state apparatus. The contemporary state is left with the almost impossible task of devising pro-commons strategies within overarching anti-commons priorities. At face value, the responses emerging from the state may seem intended to remove socionatural degradation, but the same initiatives are constrained by the ideological, symbolic and material repercussions of the private appropriation of the commons. In trying to articulate socially acceptable answers to the environmental contradictions of an unfair pattern of economic growth, the state remains reluctant (and, ultimately, unable) to represent the interests of the whole society in terms of an effective solution to environmental dilemmas. Experience shows that environmental statehood may serve to impose some constraint on private property, commodification, and accumulation, but only to the extent that it is politically acceptable to the strongest interested parties and only in order to preserve long-term socioeconomic objectives. That is why environmental politics and the rationale for environmental management remain, primarily, a struggle for the control of the state and its functional capacity (Healy, 2012). The tension between private and collective demands was captured in the long debate about the economic and political autonomy of the state as an attempt to secure the cohesion of society as a whole (Clarke, 1991). Moreover, this same debate now needs to incorporate the environmental dilemmas of the state and its role in the pursuit of core socioeconomic and political objectives.

In order to illustrate the intricacies and contradictory roles of environmental statehood, let us start with an extravagant tale about a national state trying to resolve an urgent problem. Deep in the Amazon forest, the Peruvian Government faces a serious crisis that is slowly undermining the authority of the army and affecting the morale of the troops, namely the severe lack of feminine company in the lives of the soldiers and their desperate attempts to seek relief from their anguish by sexually harassing the local female population. The problem is brought to the attention of the High Command in Lima, which was informed that "the troops in the jungle are screwing the local women.… There are rapes all over the place and the courts can't handle them all. The entire Amazon District is up in arms." After serious consideration, the generals think they have found the perfect solution: an organised and systematic provision of professional sex services through an 'off the record' agency discreetly maintained by the army. The High Command summons one of its best administrators, Captain Pantaleón Pantoja, to set up, with the necessary prudence, a 'Special Service' to assist the soldiers with their loneliness problems. To the general astonishment of his superiors, but great contentment of the local troops, Pantoja develops the most effective, punctilious and comprehensive mechanisms of mass prostitution ever seen in the jungle. Putting into practice his business management skills, Pantoja initially runs practical experiments to determine the optimal time for every 'visit' by the soldier to the female practitioner (i.e. the prostitute). The entrepreneurial captain efficiently recruits competent members of staff, develops complex timetables and supervises the intricate operation.

To everybody's bewilderment and surprise (particularly the prostitutes'), the discipline and everyday life of the Special Service begins to resemble the most austere garrison routine, with rigid codes and severe punishments. The achievements are so impressive that it quickly becomes the main military issue in the

Amazon. Soldiers, officers and even civilians increasingly talk about, and fervently demand, the assistance provided by Pantoja's diligent *visitadoras*. Tensions and disputes around the sex-delivery business keep growing, not because of bad results, but due precisely to its brilliant performance. In the words of General Victoria, one of the captain's superiors, the serious issue is not the negative things about the Service, "but the positive ones". Inadvertently, Pantoja has designed and operated the most professional and hard-working branch of the army. The story goes on (including several other amusing developments), but, as expected, it ends with the collapse of the whole enterprise. The girls continue to be extremely competent and the work is increasingly successful, but Pantoja falls prey to a beautiful Brazilian woman, ruins his marriage and, in a despairing act of love, openly exposes the association between the Special Service and the army. The public becomes aware that behind the prostitution supply lies the national state and the entire operation had been managed by army officers. Pantoja refuses to give in and, despite all the scandal, tries to re-establish and even expand the Special Service, until he is eventually relocated to a remote part of the country where he is allowed to practise his organisational skills.

This fabulous novel, written by Mario Vargas Llosa (in English, it received the title 'Captain Pantoja and the Special Service'), deals with the increasing rationalism of state solutions in response to the overarching irrationality in the assessment and conceptualisation of problems. Vargas Llosa (1987) offers a satirical phenomenology of the organisation and functioning of the Peruvian state machinery and its reaction to a challenging situation. The answer devised by the central administration in the national capital seems to bring an apparently logical solution to the problem: if there is an urgent demand for feminine company, let's deliver enough women to appease the troops and preserve the image of the army. However, the service devised by Pantoja resolves only the most immediate crisis, and ends up exacerbating existing problems and generating new controversies. Therefore, it can be treated as a metaphor for the contradictions between a sophisticated state apparatus and unresolved socioecological problems. The failure to prevent and properly deal with negative outcomes is not something associated with the lack of interest in environmental issues, but environmental management questions keep growing unabated along with the organisation of an extensive environmental statehood. In a similar way to Pantoja's sex-delivery machine, the administrative branches of the present-day state contain an extensive number of agencies and an army of environmental regulators, who are typically equipped with advanced monitoring data and bio-physical information.

As in the case of the 'visitadoras', the bounded rationality of environmental statehood systematically fails to address the root causes of environmental problems. It is obvious that the contemporary state is a complex, multilayered institutional apparatus that exercises its authority over territory and people through the implementation of some form of environmental statehood. Among the many responsibilities of the state are the containment of environmental conflicts, the remediation of impacts, the provision of essential public services and the inhibition of rent-seeking behaviours that are considered unacceptable. Moreover, to the same extent that the political agenda of elected governments is not automatically fair and democratic, extensive and sophisticated environmental regulation does not necessarily lead to ecological conservation. This is fundamentally because, instead of a commitment to resolving past issues and new tensions, the state actually operates through a combination of narrow rationalism within the wider irrationality of socioeconomic activities. The result is the perpetuation of multiple troubles associated with the access, management and conservation of socionature. The more the state tries to organise and regulate collective affairs, the more hopeless become the claims to "'rational' decision-making" (Offe, 1996: 63). As famously observed by Max Weber, rationalisation is the fate of the contemporary society, and state institutions are increasingly based on rational–legal authority.

This inherent tension between rationality and irrationality that characterises contemporary environmental statehood is closely related to, and actually derives from, the basic responsibility of the capitalist state towards private property and profit maximisation, at the expense of the commons and the social groups directly dependent on the commons for their survival. From the local to national and international spheres of environmental statehood, the apparatus of the state is increasingly required to compensate for market failures and resolve environmental disputes (the rational side of state action), but that is only achieved through the development of socioeconomic institutions associated with environmental disruption and the conversion of collective assets into private gains. This reveals the irrational or, at best, constrained rationality of the state apparatus. Enevoldsen (2001: 73) rightly declares that "the individual rationality of the actors stands out as the central ontological principle on which modern environmental regulation has been built". It means, just like in the enterprise eagerly carried out by Pantoja, that the complicated procedures laid down by state agencies are undermined by the productivist and privatist priorities of the same state (i.e. its anti-commons obligations). The result of the inbuilt irrationality that permeates the formal, legalistic rationalisation of the state is the persistence of conflicts and, ultimately, the overall trend of environmental degradation. It should be remembered that 'private' derives from the Latin word *privare*, which means depriving of access to the public sphere.

Environmental statehood needs to be understood in this context of structural contradictions and irrational rationalism that pervade the environmental practice of the capitalist state. The introduction of environmental statehood largely replicates the wider canvas of Western politics and its conflicting treatment of individual and collective interests. In that sense, the organisation of environmental statehood through narrow rationality and political control has important parallels with what Lefebvre defines as the social production of 'abstract spaces', that is to say, the homogenisation of complex, variegated spatialities that is needed for the expansion of capitalism and the affirmation of Western modernity. "Time is thus solidified and fixed within the rationality immanent to space" and, in that way, the product of historicity is an integral part of the social production of space (Lefebvre, 1991: 23). The capitalist economy requires the simplification (in socioecological and cultural terms) of social spaces for capital accumulation and, consequently, the intervention of the state plays a crucial role as the guarantor of these abstractionist procedures. The state is also a direct beneficiary of such spatial transformations, given that it enforces its authority through the production of space and associated territorial tendencies (Brenner and Elden, 2009). The environmental mediation exerted by the state is connected with the spatialisation effected by the state according to overarching political and economic pressures. In the words of Poulantzas (1978: 114), "[n]ational unity or the modern unity thereby becomes *historicity of a territory and territorialization of a history*". The contemporary state has become the main controller of the intersections between spatial and temporal matrices, which are organised and managed to serve primarily anti-commons economic demands.

Environmental statehood and its associated state-fix

This synergy between irrational rationalisation and the production of abstract spaces by the state is directly informed by the ideological, fetishised conceptualisation of 'nature' and the 'environment' that has long characterised the Western worldview. The hegemonic way of thinking in the geopolitical West, with its multiple repercussions in the rest of the world, is based on a naturalist interpretation in which there is some material continuity but also substantial discontinuities between humans and non-humans (Descola, 2005). This separation between society and the rest of nature is not only one of the main sources of environmental degradation, but it also affects the very efficacy of environmental regulation and policy-making. As pointed out by Whitehead *et al.* (2007: 20), the consolidation of a particular type of Western knowledge "has provided the ideological and technological context within which the intensive centralization and territorialization of the world (which we claim characterizes contemporary state–nature relations) has flourished". The most perverse consequence of the systematic cleavage of society from the rest of nature is the prioritisation of the needs of immediate socioeconomic relations above the long-term requirements of ecosystems and other living species. More importantly, the rationality of environmental statehood is deeply entrapped in this antagonism between the human and 'more-than-human' dimensions of the world, for example in situations where the state promotes or tolerates the over-exploitation of timber, mineral and biodiversity, receives revenues from the taxation of mass consumption and mass waste or allows the destruction of ecosystems for the construction of large infrastructure projects. Such anthropocentric priorities do not evenly reflect the wishes of all members of society, but are basically determined by the immediate demands of the most powerful and influential social groups (who are typically those in control of the state apparatus).

Instead of a passive bystander, the state has been an active player in the expansion of Western modernity and, consequently, in the reinforcement of the ideological disconnection between society and the rest of nature. The consolidation of a market-based society informed by modern thinking has extended the rate of biophysical transformations, and led to an ideological departure of society from nature and the ensuing generalisation of environmental degradation. This is because the monetisation and circulation of commodities depend on the simultaneous, and sustained, exploitation of people and nature or, as Marx (1973) put it, on the separation of humanity from its natural conditions that results from the commodification of labour and the dominance of capital. Rather than bridging the artificial gap between society and nature, the introduction of environmental statehood is firmly based on the systematic reassertion of the same dichotomist mindset. Environmental statehood had to be introduced because, when economic activities started to continuously encroach upon ecosystems and ecological features, the state was required to concoct particular techniques and procedures for dealing with environmental problems and associated political reactions. The expansion of Western modernity was not only one of the central goals of the capitalist state, but the state has been constantly modernised, albeit in a context of increasing alienation of society from itself and its own ecology. Anthropocentrism was obviously already active in the early Portuguese and other European mercantilist navigations since the fifteenth century, but with technological and economic intensification it provided the necessary justification for the conversion, since the mid-twentieth century, of nature conservation into an 'accumulation strategy' in itself (Katz, 1998). It led Latour (2004: 58) to conclude that the

> historical importance of ecological crises stems not from a new concern with nature but, on the contrary, from the impossibility of continuing to imagine politics on one side and, on the other, a nature that would serve politics simultaneously as a standard, a foil, a reserve, a resource, and a public dumping ground.

Therefore, a politico-ecological perspective needs to commence from a serious critique of the dualist rationale of Western modernity – endorsed, promoted and used by the state – that ends up negating socioecological consciousness and, in the process, denies nature its own protagonism. What is commonly described as 'nature' in effect comprises a wide range of 'socionatural assemblages'; intertwined, perpetual interrelationships between the (crudely defined) 'social' and the 'natural' (Asher and Ojeda, 2009). Instead of a static, detached category, 'nature' is also an important agent in historical and geographical change. It means that agency is effectively shared between society and the rest of nature, in sharp contrast to conventional interpretations that ideologically locate historico-geographical initiatives exclusively in the social realm. Wrongly conceptualising it as emptied of agency, it becomes much easier to conquer and control socionature and minimise the resistance by individuals and grassroots organisations affected by socioecological disruption. Western modernity mistakenly erects barriers in a world where, in fact, there are no fixed boundaries separating the human from the 'more-than-human' dimensions. The reality of the world is, rather, intrinsically socionatural, that is, it is a hybrid category that embodies processes that are simultaneously material, discursive and symbolic (Swyngedouw, 2004). As argued by Williams (2005: 81), when

> nature is separated out from the activities of men, it even ceases to be nature, in any full and effective sense. Men come to project on to nature their own unacknowledged activities and consequences.... Ideas of nature, but these are the projected ideas of men.

Yet there is a persistent difficulty among most environmental 'experts' and policy-makers trained in the modern Western tradition to see the wood for the trees and comprehend the profound importance of a socionatural epistemology and its related vocabulary. The advance of environmental statehood has basically followed the anti-commons imperatives of the state and produced only hesitant attempts to handle the accelerating disruption of socionature. Consequently, environmental regulation and policy-making have been decisively contained within the dichotomic thinking of Western modernity and the supposed rationality of broader socioeconomic irrationalities. Similarly to Pantoja's folly in the Peruvian jungle, more comprehensive environmental policies have been adopted in the last five decades, but most strategies have been systematically weakened by the teleology of environmental statehood. Notwithstanding a discourse in favour of integration, participation and prevention, environmental statehood's main purpose has been to soften and, ultimately, justify exploitative processes that depend, directly or indirectly, on the appropriation of socionature for private economic gains. Although there are cases where the results are considered broadly positive and environmental protection is successful (such as the international moratorium on whale hunting agreed in 1982 and the response to the ozone hole through the 1987 Montreal Protocol), in most situations environmental statehood has been reactionary and fragmentary. In addition, the limited prospects of environmental statehood have been further aggravated by the restricted amount of technical information and the common inadequacy of funding, regulation and staff involved in environmental management and regulation.

Although environmental degradation is not new in historical terms, environmentalism and environmental politics achieved a higher level of importance with the spread of urban–industrial capitalism since the late nineteenth century. The need to address environmental questions triggered more integrated assessments and policy-making required to deal specifically with the socioecological impacts of intensified production. The environmental question has since then been increasingly connected with other sectoral policies on transport, housing, food production, public health, etc. In practice, however, despite the colourful language used by regulators and politicians, the central reason for introducing environmental statehood has been the diminution of the most controversial troubles and the containment of protests formulated by those more directly affected by the disruptive management of resources and ecosystems. The evolution of environmental statehood has been characterised by this perennial tension between rhetorical intentions, contingent responses and multiple forms of political contestation. In actual fact, the conservation and restoration objectives of environmental statehood have been typically hindered by the wider political game and the more powerful socioeconomic demands. Not only that, but the violence perpetrated by the state apparatus on behalf of economic development and public order has also marked the trajectory of environmental statehood. Similar to what Marx affirmed in relation to the expanding bourgeois state of his time, the effects of environmental statehood, because of the basis of capitalist society, become dissociated from the living consequences of environmental degradation and the actual hardship suffered by marginalised individuals and groups. Marx expressed frustration with the outcome of the French Revolution and the national state that resulted from the manipulation of the popular upheaval by the emerging bourgeois class. In his plan for a book specifically about the state, Marx (1844) schematically noted "[t]he self-conceit of the political sphere – to mistake itself for the ardent state.... All elements exist in duplicate form, as civil elements and [those of] the state." In the case of our current investigation, the components of environmental statehood exist at a distance from the fundamental causes of socioecological degradation and, what is worse, serve to delay the construction of 'ardent' (i.e. effective) solutions.

All this makes environmental statehood only a temporary politico-institutional settlement aimed at providing, at best, partial answers to only the most urgent environmental problems. The recent history of the state clearly demonstrates that the great majority of its environmental responses have been circumstantial and have observed utilitarian objectives. The agencies, norms and procedures adopted by the state reflect the reactive and conditional basis of environmental statehood. Consequently, the operational components of environmental statehood are malleable, dynamic and function for a certain given time as a state-fix. Each state-fix includes a range of transient mechanisms and regulatory instruments organised according to political compromises, social demands and the ability of the state to respond to the most pressing environmental questions. While environmental statehood encapsulates the rationality and political thinking of the historical moment, the state-fix contains the more recognisable tools and procedures of the state for dealing with environmental issues. The need to have state-fixes in charge of the daily management of socioecological problems corresponds to the observation that "every form of production creates its own legal relations, form of government, etc." (Marx, 1973: 88). It is by definition a 'fix' because of the inability of the state apparatus to detach itself from the socionatural contradictions of capitalist economy and produce long-term responses to the intrinsic need to systematically exploit society and the rest of socionature.[1]

Similarly to the introduction of environmental statehood, the organisation of new state-fixes has played an active role in the reconfiguration of the state apparatus. Such environmental statecraft has not followed a linear renovation pattern, but the relative stability of a given politico-institutional arrangement has tended to eventually be disrupted because of new environmental problems and, also, the more general reforms of the state apparatus. In general terms, the history of environmental statehood – with the corresponding forms of state-fix – can be divided into three main periods: the early model, the conventional model and the neoliberal model. It is not the intention of this book to present a comprehensive review of the details and functioning of each model, but our main objective is to explore the historico-geographical and politico-philosophical foundations of the different state-fixes. The early statehood lasted from the turn of the twentieth century to the late 1950s, and its most relevant purpose was to regulate the access and allocation of natural resources and offer some initial form of pollution control (particularly in terms of river and water degradation). In this phase, the first examples of environmental legislation were passed, and proto-state agencies were established for dealing with ecological impacts and the management of territorial resources. The limitations of this early model became increasingly evident and it was replaced by the conventional framework of environmental statehood that lasted from the 1960s to the 1990s (with some evident repercussions and legacies beyond this period). In this phase a series of dedicated environmental regulatory agencies were formed, detailed environmental legislation was approved and the environmental question became part of public policy-making.

However, the high costs and the modest results of the conventional framework were increasingly condemned by different sectors of society and, on a par with the neoliberal reform of the state apparatus, more flexible, and market-friendly approaches were put in place from the 1990s. This last and ongoing phase of environmental statehood has been enthusiastically celebrated by regulators, politicians and policy-makers – *à la* Pantoja – particularly because of the alleged progress towards ecological modernisation, sustainable development and environmental governance. For those players in charge of the state apparatus, the current configuration of environmental statehood represents the most sophisticated and effective response possible to challenging environmental problems (vis-à-vis, for example, the rhetoric of the environmental agencies of the European Union and OECD countries). On the other hand, flexible environmental statehood has also provided favourable opportunities for the expansion of the neoliberal agenda using socionatural systems and environmental management. In the last two decades, the configuration of environmental statehood has been adjusted to the needs of a global-market society and, as a result, environmental conservation has become more closely connected with the circulation and accumulation of capital (rather than with the more common restrictions on productive and extractive activities that characterised the conventional model of environmental statehood). In the absence of explicit market mechanisms, flexible environmental statehood has given rise to 'market proxies', as in the case of bulk water charges, utility privatisation and the payment for ecosystem services. Nonetheless, in most cases, neoliberal environmental policies have aggravated social differences and created additional competition for territorialised resources and the global commons. The many tensions associated with the neoliberalisation of nature have often led to questionable operational results, greater anxiety and increasing political disputes, which ultimately affect environmental conservation and undermine policy objectives.

It is relevant to note that the environmental statehood currently in place in the early twenty-first century combines many elements of the previous, conventional phase. For instance, in many countries the reactions to global climatic change have merged conventional rules and penalties with the introduction of economic incentives and new market transactions of neoliberal inspiration. This is because the transition from one model of environmental statehood to the next is not complete, or decontextualised. In order to understand the political and social complexity of present-day environmental statehood (and its associated state-fix), it is important to realise that the actual implementation of environmental statehood is highly reflective of the particular historical and geographical circumstances of the state apparatus. Environmental statehood contains centralised and diffuse elements of authority in conformity with the specific features of politicised relations and socioecological pressures. As observed by Jessop (1982), the state is an institutional ensemble of power centres that does not exist in isolation from the balance of political forces, but these forces are in fact responsible for shaping – at least in part – the structure and intervention of the state. That is why there is never a full correspondence between the environmental responses of the capitalist state and the socioecological interests of the dominant classes, and environmental statehood is therefore constantly contested and renegotiated.

This operational and political selectivity of the environmental statehood is not established in advance, but it is the result of the interplay between state priorities and sociopolitical contestation within and beyond state institutions. As pointed out by Swyngedouw and Heynen (2003: 912–13),

> socioecological processes give rise to scalar forms of organisation – such as states, local governments, interstate arrangements and the like – and a nested set of related and interacting socioecological spatial scales.... These territorial and networked spatial scales are never set, but are perpetually disputed, redefined, reconstituted and restructured in terms of their extent, content, relative importance and interrelations.

In order to capture the intricacy of the responses formulated by the state regarding multiple environmental problems, it is necessary to go beyond both conventional techno-bureaucratic explanations and superficial attempts to integrate state, society and the wider socionature. It is not enough for a political ecology of the state to replicate such mechanical connections that, incidentally, permeate the ordinary calls for sustainable development, ecological modernisation and environmental governance. The state apparatus needs to be understood in relation with, and connected to, the interdependencies and contradictions between society and the rest of socionature. This is the last main topic to be examined in this chapter.

The trialectics between state, society and the wider socionature

A genuine politico-ecological perspective of the state should neither accept a simplistic, apolitical consideration of the 'more-than-human' domain, nor conceive of nature as the passive recipient of social agency (as in most initiatives informed by Western modernity). The shortcomings of a crude socioecological reasoning are nothing but the containment of environmental politics within anthropocentric boundaries and non-critical postulations. As an attempt to fully integrate interconnected sub-systems, it is well known that Hegel used the concept of sublation [*Aufhebung*] in his quest to settle any opposition between the individual and his or her collectivity. Sublation is the process in which contradictory positions are reconciled in a higher synthesis where they are both preserved and changed through a dialectical interplay. For Hegel (1977), in contrast to simplistic bifurcations between the pedestrian apprehension of right and wrong, there are multiple, intermediary conditions in the reality of the world (which is something extremely relevant today regarding the lingering dilemma between group demands and domineering state power). Lefebvre (1991) critically expanded Hegel's argument by observing that dialectical thinking requires a better and more organic inclusion of the 'third term'. This means the rejection of a closed logic of 'either/or' in favour of the radically open formulation of 'both/ and also'. According to Lefebvre, in order to overcome the enduring problems of contemporary society, there should always exist the possibility of the 'other' and the opportunity to reach unforeseen, creative outcomes. Beyond the static dichotomies of Western modernity, 'thirdering' assumes crucial importance, that is, in a dialectical trinity (*trialectics*), where the third term resolves, and reopens, the tension between the previous two.

Lefebvre (1968) believed that the third term – not merely the resultant of the dialectics of thesis/antithesis and synthesis, but as something equivalent to, and interrelated with the other two – can become the solution to the problems of contemporary conflicts and contradictions. If the dialectical approach is to transcend formal logic, the recognition of a more 'embellished' third term allows for a fruitful reunion of opposing positions without losing the invigorating force of the opposition. 'Thirdering', or the acceptance of an inclusive trialectical continuum, can represent a radical epistemology that produces new alternatives that are both similar and different (Soja, 1996). In our case, the state is obviously an integral element of socionature that exists because of the co-evolution, and interdependencies, between the social and the 'more-than-social' realms. But at the same time the state has its own specificities that emerge depending on the concrete characteristics of socionature, and the state's attempts to interfere with it. Following Lefebvre, the state is a socionatural driving-force, but it also takes some distance in order to mediate and regulate socionatural processes. The special 'function' or 'position' occupied by the state is not detached from socionature, but takes place through trialectical interactions. Through a trialectical ontology of the interconnections between state, society and the rest of nature, it is easier to appreciate the uniqueness of the state as a moment, a particular locus within the socionatural world. The three categories that form the trialectics – state, society and the rest of nature – are not separate, but constitute three inherently connected factors of the same relational, contested reality.

The removal of the binary, inflexible dialectical interpretations of historical materialism and its replacement with the more creative trinitarian formulation have serious consequences for political ecology and environmental politics. A trialectical ontology can play an important role in the rejection of the common Western dichotomy between nature and society, as well as of the dualism between state and society. The real meaning of 'thirdering' is to yield a dynamic conceptualisation that refers to the interdependencies and contradictions between social groups, their socionatural condition and the possibilities of state administration. The strategic and relational understanding of state action can be reformulated as a 'trialectical' relationship between society and nature with the state as the emerging 'third term'. The third term – in the case of the evolution of environmental management, the state – is not the simple outcome of the interaction of the other two, but corresponds to a contested locus of action that reshapes, and simultaneously evolves, with nature and society. As mentioned above, in the history of environmental statehood, the state constitutes the mediator and executor of social demands, according to the balance of power and socioecological conditions, but a trialectical ontology can now help to embrace these dynamic and politicised interactions between socionature and the state. Finally, the recognition of the trialectical basis of environmental management permits a better representation of the convergences and antagonisms between state and society that affect, and are affected by, environmental change.

The description of triads is certainly not new in the Marxist conception of the world. For example, Marx mentioned the internal contradictions of capitalism as the interplay between labour, capital and land, just as Gramsci saw the state as the vertex of a trialectical structure between state–economy–civil society and Jessop makes reference to the triplet market–state–society. However, the explicit consideration of the state as the 'third term' of the interplay between society and the rest of nature offers a more creative and inclusive understanding of environmental management themes. Looking from this renewed perspective, environmental problems are grounded in the evolution of the state apparatus (as a semi-autonomous phenomenon) in relation to class-based struggles (including classes, groups and interpersonal connections) and to socionatural elements. The state is the third term that both reinforces the powers of society over nature and, in an attempt to secure social cohesion, regulates the rate of socioecological impacts. In other words, the action of the state, as the third term trialectically interrelated with social classes and metabolised nature, does not necessarily favour the interests of the stronger social groups, but is the quintessential focus of contradictions, bargaining and contestation. Following the trialectical reinterpretation, the failures of environmental management can be translated as the product of the uneven balance of power within the intricate relationship between society and the rest of nature mediated, and reshaped, by the state. Power is not limited to the realm of the state and traditional politics, but permeates the channels of interaction between society and historicised nature. Furthermore, trialectics should not be seen as an artificial ontological representation of reality aimed to create artificial boundaries, but as the vivid incorporation of highly interconnected categories that extend political action to the 'more-than-human' spheres of the world (in the sense that nature is not a passive entity that shall be dominated and exploited, but the transformation of nature reflects back to society and state).

Instead of a pre-ordained, rational structure, the state constitutes a multifaceted, contradictory and multiscalar organisation, and its evolution reflects the politicised connections between society and the rest of nature. No state apparatus exists in a socioecological vacuum, just as there is no nature outside of history. The state co-evolves with nature and society in a truly trialectical relationship in which the processes of change are simultaneously imprinted on nature, society and the state.[2] The state, as the third term of the trialectics, is not the simple outcome of the interaction between the other two, but corresponds to a contested locus of action that reshapes, and simultaneously changes with, nature and society. The state is not an entity merely derived from class relationships, but it is a relatively autonomous player with its own form, functions and apparatus (Clark and Dear, 1984). The prominent role of the state apparatus in the trialectics between society and the rest of nature is not something that happens by chance. On the contrary, as in the case of the responses formulated by the contemporary state, environmental statehood is an evolving compromise between opposing forms of socionatural interaction and their conflicting justification and rationalities. Since the early legislation concerned with pollution control at the end of the nineteenth century, the environmental regulatory functions of the state have had this contradictory character, that is to say, the difficulty in reconciling, in the same regulatory instrument (such as laws, norms and programmes), highly antagonistic demands. In practice, the main role of environmental statehood has been the sanctioning of 'non-commonalities', that is, the consolidation of the transformation of the commons (resources and ecosystems shared by individuals and communities) into proxies or legally recognised forms of private property (the non-commons).

The intention of the elaboration presented in the previous pages – including the brief discussion on rationality, state-fix and trialectics – is to make clear that the tension, continuity and creativity around environmental issues is primarily located in the constraints and in the possibilities of the capitalist state (which is in itself the incarnation of contradictory rationalities, demands and propositions). Power exercised by the state abolishes the distinction between animate and inanimate, inert and living, before and after (Lefebvre, 2009). As a result, the state cannot be understood as merely the controller of settled environmental protection principles, but it is itself a central source of conflicts and compromises. In the Marxist–Gramscian sense, the state is more than just the apparatus of government and policy-making, but it contains the contested association between political society and civil society. Gramsci brings attention to the 'integral state', a fluid, dialectical unity that encapsulates a complex web of social relations (Nash, 2013). The state is a continuous process of "formation and superseding of unstable equilibria [...] between the interests of the fundamental group and those of the subordinate groups" (Gramsci, 1971: 182). The state is the political society that rules a nation, but it is also a circumstantial settlement between political society and civil society that follows the hegemony of power (Gramsci, 2011). The state apparatus (which Lefebvre differentiates from the 'state itself') contains not only a collection of agencies and legal instruments, but is essentially a complex of structures and strategies that reflect the balance of political power and social antagonisms (Lefebvre, 2008). Instead of a preordained, rational entity, the state constitutes a multifaceted, contradictory and multiscalar organisation, and its historico-geographical evolution reflects the politicised connections between society and the rest of socionature.

The various concepts discussed here help to focus the debate on the interconnected exploitation of society and the rest of nature that has defined the expansion, and the vitality, of socioeconomic measures advanced by the contemporary (capitalist) state. The state–nature–society nexus exposes the systematic attempt of the contemporary state to sustain the logic of exploitation, alienation and accumulation. The analysis of the political ecology of the state should depart from Hegel's politico-philosophical thinking and his narrow dialectical epistemology. It should take a critical position against the Hegelian state theory based on the amelioration and consolidation of bourgeois society, which are processes that benefit from a renovated, but inherently conservative, state apparatus. The contemporary (capitalist) state is unable to handle environmental matters beyond the hegemonic relations of production and reproduction. The command over nature, as famously announced by Francis Bacon, has mutated into contemporary environmental regulation not only as an additional task, but as an integral part of the justification and vibrancy of the capitalist state. Therefore, a class-based critique should start with the consideration of the state as not simply an external, distant controller of environmental questions, but with an understanding of socionatural relations as a crucial factor in the organisation and functioning of the state.

The current chapter has served to clarify, through an eco-Marxist interpretation of the environmental statehood, many important factors that guide and limit state interventions. Such an approach can help to examine the role of the environmental question in the development of capitalism and, in particular, the unrelenting, but capricious, transformation of the commons into marketable private property. The state, in charge of capitalist relations of production, has been expected to preserve economic liberties as long as they leave the asymmetric distribution of private property and socioeconomic opportunities unaffected. The next chapters will analyse the formation and transition of environmental statehood, highlighting the conceptual premises of the different models and their fundamental shortcomings. In doing this, the other parts of this book will show that, although environmental statehood as such was only adopted in the last hundred years or so, its rationalisation, organisation and failures echo the wider debate about the role, and the constraints, of the state through the expansion of capitalist relations of production and reproduction. If the modern state is the result of the politicised exchanges between different social groups and the rest of socionature, environmental statehood is only one of the most recent, better defined responsibilities of the state that is necessary in order to deal with the ecological contradictions of capitalist encroachment upon socionature.

The dynamic evolution and the critique of environmental statehood betray traces of the ideas advanced by Hobbes (e.g. the justification of a powerful authority above the rest of society and the production of private property by the state), Kant (e.g. the enlightenment of the social contract according to calls for freedom and reason, the social purpose of private property, and a harmonious relation between state and society), Hegel (e.g. the formation of flexible, legitimate state arrangements that include executive and legislative branches of the state under the command of a supreme ruler) and Marx (e.g. the need to profoundly rebuild the state in a society freed of the disruptive effect of commodification, accumulation and alienation). The reference to those paradigmatic thinkers goes beyond their personal intellectual contribution, but also encapsulates the long-term legacy of their ideas which re-emerged, and are still very much in evidence, in the history and practice of environmental regulation. While Hobbes conceived of the state as the guarantor of social norms and market rules, against differing, primitive wishes of the individuals, Hegel crafted the most sophisticated and comprehensive explanation of the role and necessity of the state, which provided the basis for the more recent and flexible model of environmental statehood. From a critical perspective, Marx (1975: 217) advised that "[h]istory has been resolved into superstition for long enough. We are now resolving superstition into history." Such emblematic philosophers have not only influenced policy-making and the organisation of the state, but their ideas also condensed the controversies around the appropriation of the commons and the intrinsic difficulties the capitalist state faces when mediating socionatural controversies.

Notes

1 Although environmental statehood observes the mainstream preservation of capitalist relations, the negotiations around the formation of a state-fix can sometimes lead to compensations and concessions to some affected groups (beyond the more usual and predictable mitigations that are typically consistent with the asymmetric balance of power). Undoubtedly, these concessions are normally secured through the reform of the state apparatus rather than directly from capital, but nonetheless they still represent an 'extra' economic cost to the propertied classes. These are typically isolated developments (e.g. the momentary approval of more stringent environmental legislation or a more strict enforcement of norms and rules) that are soon accommodated within the anti-commons commitments of the state and the subordinate, adjunct role of environmental statehood.

2 At this point in this text it should become clear that our use of the words 'society', 'nature' and 'state' is more didactic than ontological given that these are trialectically related categories.

Bibliography

Agnew, J.A. 2001. Disputing the Nature of the International in Political Geography: The Hettner-Lecture in Human Geography. *Geographische Zeitschrift*, 89, 1–16.

Asher, K. and Ojeda, D. 2009. Producing Nature and Making the State: Ordenamiento Territorial in the Pacific Lowlands of Colombia. *Geoforum*, 40(3), 292–302.

Brand, U. and Görg, C. 2008. Post-Fordist Governance of Nature: The Internationalization of the State and the Case of Genetic Resources – a Neo-Poulantzian Perspective. *Review of International Political Economy*, 15(2), 567–89.

Brenner, N. 2004. *New State Spaces: Urban Governance and the Rescaling of Statehood*. Oxford University Press: Oxford.

Brenner, N. and Elden, S. 2009. Henri Lefebvre on State, Space, Territory. *International Political Sociology*, 3, 353–77.

Clark, G.L. and Dear, M.J. 1984. *State Apparatus: Structures and Languages of Legitimacy*. Allen and Unwin: Winchester, MA.

Clarke, S. (ed.). 1991. *The State Debate*. Macmillan: Basingstoke and London.

Descola, P. 2005. On Anthropological Knowledge. *Social Anthropology*, 13(1), 65–73.

Enevoldsen, M. 2001. Rationality, Institutions and Environmental Governance. In: *Environmental Regulation and Rationality: Multidisciplinarity Perspectives*, Beckmann, S.C. and Madsen, E.K. (eds). Aarhus University Press: Aarhus, pp. 73–111.

Engels, F. 2010. *The Origin of the Family, Private Property and the State*. Penguin: London.

Gerstenberger, H. 2011. The Historical Constitution of the Political Forms of Capitalism. *Antipode*, 43(1), 60–86.

Gramsci, A. 1971. *Selections from the Prison Notebooks*. Trans. Q. Hoare and G.N. Smith. Laurence and Wishart: London.

Gramsci, A. 2011. *Lettere dal Carcere*. Einaudi: Torino.

Harvey, D. 2006 [1982]. *The Limits to Capital*. Verso: London and New York.

Healy, K. 2012. Sociology. In: *A Companion to Contemporary Political Philosophy*, Goodin, R.E., Pettit, P. and Pogge, T. (eds). Wiley-Blackwell: Oxford, pp. 88–117.

Hegel, G.W.F. 1977 [1807]. *Phenomenology of Spirit*. Trans. A.V. Miller. Oxford University Press: Oxford.

Jessop, B. 1982. *The Capitalist State: Marxist Theories and Methods*. Blackwell: Oxford.

Katz, C. 1998. Whose Nature, Whose Culture? Private Productions of Space and the Preservation of Nature. In: *Remaking Reality: Nature at the Millennium*, Braun, B. and Castree, N. (eds). Routledge: New York, pp. 46–63.

Latour, B. 2004. *Politics of Nature: How to Bring the Sciences into Democracy*. Trans. C. Porter. Harvard University Press: Cambridge, Mass. and London.

Lefebvre, H. 1968. *Dialectical Materialism*. Trans. J. Sturrock. Cape Editions: London.

Lefebvre, H. 1991. *The Production of Space*. Trans. D. Nicholson-Smith. Blackwell Publishing: Oxford.

Lefebvre, H. 2008. *Space, Difference, Everyday Life*, Goonewardena, K., Kipfer, S., Milgrom, R. and Schmid, C. (eds). Routledge: New York.

Lefebvre, H. 2009. *State, Space, World: Selected Essays*. Trans. Moore, G., Brenner, N. and Elden, S. University of Minnesota Press: Minneapolis, MN.

Marx, K. 1844. *Draft Plan for a Work on The Modern State*. MECW Vol. 4, p. 666. Available at www.marxists.org/archive/marx/works/1844/11/state.htm

Marx, K. 1973 [1857–8]. *Grundrisse*. Trans. M. Nicolaus. Penguin and New Left Review: London.

Marx, K. 1975. *Early Writings*. Trans. R. Livingstone and G. Benton. Penguin and New Left Review: London.

Nash, F. 2013. Participation and Passive Revolution: The Reproduction of Neoliberal Water Governance Mechanisms in Durban, South Africa. *Antipode*, 45(1), 101–20.

Offe, C. 1996. *Modernity and the State: East, West*. Polity Press: Cambridge.

Poulantzas, N. 1978. *State, Power, Socialism*. Trans. P. Camiller. Verso: London and New York.

Soja, E.W. 1996. *Thirdspace: Journeys to Los Angeles and Other Real-and-Imagined Places*. Blackwell: Cambridge, MA and Oxford.

Swyngedouw, E. 2004. *Social Power and the Urbanization of Water: Flows of Power*. Oxford Geographical and Environmental Studies. Oxford University Press: Oxford.

Swyngedouw, E. and Heynen, N.C. 2003. Urban Political Ecology, Justice and the Politics of Scale. *Antipode*, 35(5), 898–918.

Vargas Llosa, M. 1987. *Captain Pantoja and the Special Service*. Trans. G. Kolovakos and R. Christ. Faber and Faber: London and Boston.

Whitehead, M., Jones, R. and Jones, M. 2007. *The Nature of the State: Excavating the Political Ecologies of the Modern State*. Oxford Geographical and Environmental Studies. OUP: Oxford.

Williams, R. 2005 [1980]. *Culture and Materialism*. Verso: London and New York.

Wissen, W. 2009. Contested Terrains: Politics of Scale, the National State and Struggles for the Control over Nature. *Review of International Political Economy*, 16(5), 883–906.

3 Early and conventional environmental statehood

Unpacking the Hobbesian influence

The rise of environmental statehood

The consolidation of urban–industrial capitalism in the Western world since the middle of the nineteenth century – based on the intensification of manufacturing, the commodification of labour-power, and the conquering of new territories and resources around the globe – meant an even more systematic encroachment upon socionature as a means to the accumulation of wealth. One of the hallmarks of the urban–industrial socioeconomy was its peculiar and disturbing transformation of ecological and social features beyond the resilience of socionatural systems. This increasing economic activity and fast technological innovation resulted in the by-product of generalised forms of environmental impacts and the associated forms of protest. Problems such as water pollution, deforestation and biodiversity loss started to draw increasing attention from governments and early environmental activists. Artists and intellectuals, such as Henry David Thoreau, John Muir, George Perkins Marsh, William Morris and John Ruskin, helped to somehow increase public awareness of environmental matters and open them to wider debate. In an anticipation of the politico-ecological concerns that emerged one century later, Marx and Engels also argued that the

> 'essence' of the freshwater fish is the water of a river. But the latter ceases to be the 'essence' of the fish and is no longer a suitable medium of existence as soon as the river is made to serve industry, as soon as it is polluted by dyes and other waste products, and navigated by steamboats, or as soon as its water is diverted into canals where simple drainage can deprive the fish of its medium of existence.
>
> (quoted in Foster, 2002: 59)

With the onset of mounting rates of environmental disruption, many scholars and political leaders understood that economic growth, public health and quality of life were becoming gradually endangered. The crux of the matter was how to allocate responsibilities and find appropriate responses to problems that were the result of private interventions (backed by the state) and which were starting to affect society as a whole, calling for specific state interference. Rising pollution and resource scarcity forced the organisation of the first, tentative attempts to regulate the appropriation and use of territorial resources and the disposal of waste. However, the underlying aim of the emerging mechanisms of environmental statehood was not the resolution of the environmental problems per se, but to sustain the long-term prospects of the prevailing anti-commons tendencies, as well as to inhibit criticism and the consideration of more radical alternatives by those affected by socioecological disruption. From its early beginnings, environmental statehood was translated into a range of legislation, policies and state agencies – corresponding to the operational side of environmental statehood, that is, the state-fix – aimed at controlling, but also facilitating, the access to socionatural systems. The contradictions of the expansion of the capitalist economy over society and the rest of nature brought about this dual, conflicting purpose of environmental statehood: on the one hand, the state remained the champion of capitalist relations of production; on the other hand, because of changing socionatural conditions, state action had to be broadened in order to mitigate growing tensions and prevent situations of scarcity or overprice of natural resources. As pointed out by Wells (1981: 67), the "secret of the state lies not within itself, but in what appears to be outside it".

It is evident that previous capitalist state formations, such as those established under the Treaty of Westphalia and the resulting "pan-European inter-state system" in the seventeenth century (Arrighi, 1994: 36) had already imposed similar sets of institutions related to the control over socionature (in Europe as well as in its colonies) that played an important economic role during the pre-industrial phase of capitalist expansion. The absolutist regimes of the time were in a unique position to advance the mercantilist economy and protect the wealthier groups' interests in the exploitation of overseas resources and the expansion of foreign markets (Poggi, 1978). At the same time, the countries with more developed capitalist relations (Holland and England at this time) had a clear advantage in the process of inter-state competition, especially because of the capacity of those states to fund and organise economic activities in different parts of the world (Callinicos, 2007). The handling of socionature by the apparatus of the state in the early capitalist world – under the priorities of mercantilism and the conquering of territorial resources – was substantially distinct from the equivalent approaches adopted by the medieval state. The majority of the European population during the Middle Ages consisted of peasants who relied on the agricultural potential of the proprietor's land and used natural resources mainly for family consumption. Society was formed essentially of peasants (vassals) and landowners (suzerains with land as the main source of power and authority). Instead of the bourgeois ideology that was later able to transform social relations into 'relations between things' (Marx, 1976), medieval morality was based on the prolonged dependency of the peasants and other social groups on the property of the feudal lord. Rather than the anti-commons character of private property under capitalism, medieval state institutions were centred on the attachment and sharing of land according to the command of a territorial ruler.

The expansion of capitalism since the Renaissance meant that the trialectical relationship of state and society with the rest of socionature, particularly in

Western Europe, gradually lost most of its medieval background of mysticism, religiosity and attachment to land. The state in the first centuries of mercantile capitalism became responsible for fostering new administrative techniques within large empires, the suppression of reactions by the social groups adversely affected (such as African and Amerindian slaves) and the minimisation of private risks associated with the colonisation of new territories (for example, by sending troops and establishing international treaties). The evolving state was put in charge of connecting territories around the world and imposing a new, mercantilist order over people in different regions and their socioecological conditions. The 'mastery of nature' – with all its complexity and elusiveness, as discussed by Leiss (1994) – could never be achieved without the power of the state over communication, transport and the territory. Informed by the outcomes of the Scientific Revolution (promoted by the state and the emerging capitalist classes), the whole planet became the object of conquest and commodification. Interestingly, as an anticipation of environmental statehood, some proto-environmental legislation was already adopted for the protection of forests and other resources for military purposes, such as the codes passed by the Portuguese Crown in the seventeenth and eighteenth centuries to restrain deforestation in Brazil and secure the supply of timber to the royal navy. Similar regulatory mechanisms were adopted by the Colbert, minister of Louis XIV and father of French mercantilism, to organise the use of water and forests in France and, a century later, in colonies such as Mauritius (Pádua, 2002).

These primitive forms of environmental intervention and nature control by the mercantile state underwent a significant, qualitative improvement in the latter part of the nineteenth century. The contemporary (capitalist) state, born in Westphalia and reconfirmed in the turbulent interlude between the US Independence, the French Revolution and the Holy Alliance, then became the main guarantor of the anti-commons strategies required for the consolidation of urban–industrial relations of production and reproduction. While the bucolic aesthetics of Romanticism were a last desperate attempt to preserve the symbolic value of the fast disappearing commons, the dominant forms of subjectivity and knowledge in the new world were increasingly associated with the morality and material demands of an urban–industrial society. In this context, the state paved the institutional road for the administration of environmental degradation according to the ideological, political and economic goals of a fast-expanding capitalist economy. Against this background of state modernisation, urban–industrial priorities and noticeable environmental disruption, a series of policies and legislations specifically dedicated to ecological issues was introduced by the main Western states at the turn of the twentieth century. The environmental regulation organised at the time of the Second Industrial Revolution (also described as the Technological Revolution, which took place between the 1880s and 1920s and is epitomised by Fordist and Taylorist mass-production methods) incorporated novel institutional mechanisms for the containment of environmental problems, but evidently within the boundaries of the hegemonic politico-economic setting.

The *early phase of environmental statehood* – as a political and institutional compromise established in this period, particularly after the First World War – was the result of a growing awareness of the environmental contradiction and self-destructiveness of capitalist expansion in Western Europe and other parts of the world. In Russia, for instance, a series of acts were passed in the period between 1855 and 1913 in relation to land use, woods, rivers, fishery and fauna, primarily to regulate the private trade of natural resources, cultivation of woods and capture of rare animals, the protection of the environment against pollution and the exhaustion of fish resources (Krenke and Chernavskaya, 2010). Another example of mounting environmental problems and early forms of regulation can be found in the management of the Clyde River in Scotland, which was one of the main industrial areas of the world in Victorian and Edwardian times. The lower river encompasses the estuary around the Glasgow metropolitan area, where morphology and ecology were profoundly altered because of human interventions during the Industrial Revolution. Due to economic expansion, the river system was significantly modified to satisfy the needs of transcontinental trade and, especially, shipbuilding. James Deas, one of its most distinguished navigation engineers, argued in 1873 that probably for no river in Great Britain has so much been done 'by art and man's device' as for the River Clyde. Water quality gradually became a matter of serious concern, with the Clyde and its many tributaries so polluted that the City of Glasgow started to look for alternative sources of public supply, and passed various pieces of legislation to remediate environmental degradation (Ioris *et al.*, 2006).

The early model of environmental statehood put forward for dealing with the consequence of urban–industrial expansion, as in the case of Scotland, consisted of an institutional approach aimed primarily at damage limitation and containing environmental risks. The consequence of the introduction of environmental statehood was the implementation of early management, conservation and restoration measures (i.e. its corresponding state-fix), which nonetheless provided only partial, biased solutions to the trend towards nature degradation. The modest results achieved by early environmental statehood were normally put down to the poor quality of the legislation and the limited amount of data available to government officials, when in actual fact the intention was never to significantly alter prevailing socioeconomic tendencies. Early environmental statehood was an adjunct of stronger economic priorities and directly subordinate to the key political commitments of the liberal and, later, Keynesian state. The expansion of environmental statehood followed what Gramsci (1971) describes as 'passive revolution' (a phenomenon closely related to the 'war of position') through which molecular changes, coming from the top, progressively modify the pre-existing composition of forces and become the matrix of new changes. The early forms of environmental statehood essentially responded, on behalf of hegemonic economic interests, to the threats to the sustainability of capitalism, and only as a pre-emptive reaction that tried to avoid the need for a more radical reorientation of economy and society. At the same time, in a context of fast-changing interactions, the state was not only expected to perform multiple regulatory functions, but it gradually became the ultimate source of additional socionatural change. The state seemed to be an unfinished project "struggling to maintain dominance upon territories, nature, and population" (Asher and Ojeda, 2009: 300) and, in the process, it itself was transformed.

One of the most notorious examples of the evolution of the wider state apparatus together with the organisation of early environmental statehood was the establishment of the Tennessee Valley Authority (TVA) by President Roosevelt in 1933, with a legal mandate for retransforming a deprived catchment area through the power of dam engineering and water management. Nature was not simply the canvas of economic production, but the very object of economic recovery and political appeasement in that part of the United States (as in the rest of the country, for that matter). It was not by chance that rich water reserves and the poor people of the southern American states provided the perfect raw material for the revival of a stagnant capitalism during the Depression years. Despite the fact that the TVA was set up as a decentralised agency aimed at incorporating grassroots demands, it was quickly and persistently dominated by larger farmers and engineering construction lobbies.

> In its capacity as symbol, the organisation derives meaning and significance from the interpretations which others place upon it. The halo thus eagerly professed is in large measure a reflection of the needs and problems of the larger groups which require the symbol and use it.
>
> (Selznick, 2011: 19)

In practice, the operation of the TVA accelerated processes of socioecological politicisation and the imposition of a large-scale productivist ideology upon socionatural systems. Informed by experiences such as the TVA, the association between early environmental statehood and politico-economic priorities was one of the elements of the Marshall Plan, which helped to rebuild engineering infrastructure and revitalise the economy in Western Europe after the Second World War. In the Global South, similar funds were made available for the construction of dams and irrigation schemes, such as the construction of dams and perimeters of irrigation along the São Francisco River in the northeast of Brazil; here, a new agency was formed in 1948, inspired by the TVA experience, and since then large dams, public irrigation and environmental conservation projects were implemented (Ioris, 2001).

The essential aspect of the introduction of environmental statehood was the shift from the simple attempt to dominate nature to the domination of nature degradation. Nature was no longer seen as merely a source of resources, energy and territorial power, but it became a more integral element of the success of social and economic activities. Ultimately, the introduction of environmental statehood in the Western world corresponded to a transition from the Aristotelian teleology (i.e. in Aristotle's words, nature makes everything for the sake of something, for some purpose) which defined the treatment of environmental issues by the mercantile state, into the inventive reasoning of urban–industrial capitalism and the intention of the contemporary state apparatus to be both normative and procedural. However, the environmental policies formulated in the early phase of environmental statehood were systematically belated and unable to resolve the continuous increase in the rate and extent of degradation. Early environmental statehood also reproduced the constraints of the liberal and then welfare state in other areas of public administration, for instance, in terms of dealing with well-being and social security. Although the state started to assume a more direct role in the economy from the early 1930s (after the Wall Street Crash in 1929 and the resulting economic crisis), its environmental policies were largely focused only on the management of landscape change and the appropriation of territorial resources (as in the case of the TVA and its large hydraulic engineering projects). Environmental legislation was limited by economic and political constraints, meaning that it only represented a partial adjustment of governmental activity to long-term processes of economic, social and political development (Meadowcroft, 2005). Examples include the British 1943 Forest Policy and the 1949 National Parks and Access to the Countryside Act, which incorporated both provisions for ecological conservation and for the renewed exploitation of territorial resources.

While these initial forms of environmental regulation and bureaucracy were being implemented, additional exploitation of natural resources and the construction of larger infrastructure schemes were tolerated because of their alleged importance to national development. This was consistent with both the incipient environmental thinking at that juncture and, more significantly, the need to spare the economy from stringent regulatory restrictions that could affect economic growth. For wider society at the time, environmental concerns were still seen primarily as a personal, moral issue rather than state business. It was something distant and primarily of interest to scientists and intellectuals. This more individualist and sporadic reaction to environmental problems began to change because of campaigns associated with the 'New Environmentalism' movement and the publication of books such as *Silent Spring* by Rachel Carson in 1962. Existing state responses were increasingly challenged, especially because of the narrow basis of early environmental statehood. As a result, from the 1960s a new version of environmental statehood was introduced, which was characterised by more comprehensive forms of regulation and the organisation of new public agencies entirely dedicated to environmental conservation.

This led to the introduction of a new institutional model that can be described as *conventional environmental statehood.* It was the arrangement that prevailed between the 1960s and the early 1990s and entailed the enforcement of more detailed environmental legislation beyond the patchy approaches of the first half of the twentieth century.[1] Conventional environmental statehood magnified the purposive-rationality (i.e. 'means-end rationality', in the Weberian sense) of the early framework of environmental statehood through the approval of new layers of environmental bureaucracy. And yet, the rationality of conventional environmental statehood was significantly restricted, and often turned into irrationalities, due to the anti-commons and productivist imperatives of developmental policies. The most paradigmatic example of the 'irrational rationality' of conventional environmental statehood was the requirement to carry out environmental impact assessments (EIA) as part of the approval of programmes and projects. An EIA is meant to identify, evaluate and try to mitigate the biophysical, social and other relevant effects of development proposals. Normally, it is defined as "an aid to decision-making" and the basis for negotiation between regulators, developers and the interested public. In practical terms, regardless of its technical complexity (much helped by the use of computer models and satellite data), the decision-making informed by EIA remained undemocratic and centralised in the hands of powerful politicians more interested in the approval of plans and projects regardless of their associated environmental impacts.

Going back to the River Clyde, after the Second World War, the economy in the river basin faced a dramatic transformation with the continuous decline in the shipbuilding industry. The region became characterised by the social ills of appalling housing conditions, chronic overcrowding and the industrial problems resulting from a collapsing manufacturing base. Serious pollution prompted the establishment of the Clyde River Purification Board, which brought the whole river basin under a single independent regulator with specific powers to control emissions by means of legally enforceable standards. After the approval of a new Act in 1965, the Board acquired more comprehensive responsibilities regarding existing discharges to inland waters and new discharges to tidal waters. As the Clyde experience illustrates, the new environmental regulation attempted to normalise and control the ecological impact of economic activities (making use, for example, of standards of acceptable pollution, prohibitions and, since the 1960s, environmental impact assessments. It coincided with the establishment of dedicated state agencies in charge of environmental regulation and the expansion of public utilities in charge of delivering water, sewage, electricity and waste collection, which nonetheless achieved only isolated and tardy results.

The best example of the changes that followed the adoption of conventional environmental statehood was the National Environmental Policy Act (NEPA), introduced in the United States in 1970 and which was then followed by a series of complementary pieces of legislation (such as the Clean Air Act, the Clean Water Act, the Endangered Species Act and the Resource Conservation and Recovery Act). This package of comprehensive environmental legislation served to amplify the responsibilities of the federal agencies in charge of regulation, especially the US Environmental Protection Agency (EPA). Through the evolution of conventional environmental statehood, the American state tried to intervene more directly in the appropriation of resources, generation of waste and disruption of ecosystems. This was achieved particularly through the enforcement of restrictive licences and the application of penalties and sanctions for abuses and disobedience. The federal government, together with state administrations and other organisations, developed an extensive set of technological requirements and environmental standards, as well as enforcement procedures and severe penalties, where necessary, for dealing with environmental problems. Underlying the implementation of conventional environmental statehood in the

USA, there was a strong understanding that the national state had the cognitive capacity and practical means to determine environmental goals and how they should be achieved (Fiorino, 2006). However, a myriad of problems associated with the conventional model of environmental statehood became increasingly evident in the United States, particularly in terms of its complexity, cost and lack of regulatory clarity (Manheim, 2009).

One of the main tools of conventional environmental statehood in the USA was the internalisation of environmental externalities, such as the cost of air pollution and soil erosion, by the economic agents. The economic value of the externality can be associated with monetary figures and captured (i.e. internalised through the polluter-pays principle) by the economic agents in the form of environmental levies or production costs. For instance, the 1970 Clean Air Act required polluters to comply with environmental standards at their own expense. Nonetheless, the very definition of 'externalities' betrays a profoundly ideological vision of the world, given that environmental degradation may be causing damage outside a private property, but it is completely contained within the economic system. "Far from representing an exceptional example of inefficiency, externalities emerge as efficient methods of pursuing the true goal of the capitalist economic system, namely profit and capital accumulation" (Panayotakis, 2011: 103). It is not difficult to perceive that this form of environmental regulation only makes sense in terms of private property, private human action and private gains. It presupposes private property and helps to expand privatist institutions over socionature and into the areas of the world not yet fully dominated by capitalist relations. The mitigation of externalities is nothing less than a demonstration of the anti-commons foundations of conventional environmental statehood. As indicated by Marx, private property is essentially an intersubjective process of interrelation between man and nature, which is directly associated with mechanisms of political domination and the reinforcement of social inequalities (Giannotti, 2011).

Also in the context of the European Union, environmental policy-making until the 1960s was low on the list of politicians' priorities and was still focused on disjointed forms of regulation with merely secondary environmental consequences. That changed slightly after 1972, when specific environmental protection measures were introduced, such as the 1979 Birds Directive (Wurzel, 2008), although these were not always legally binding and scarcely observed. Reflecting upon the European, the American and similar experiences, it can be inferred that conventional environmental statehood between the 1960s and the 1990s only partly addressed, and failed to resolve, the mounting problems of an increasingly complex and rapidly globalising socioeconomy. As in the previous phase, the overall results were far from satisfactory and, in addition, conventional environmental regulation was seen as overly bureaucratic, very expensive and highly intrusive to economic agents. The formulation of responses typically relied on technical experts and hierarchical control of environmental regulation and management. It normally entailed "technocratic and interventionist forms of top-down policy making where uniform and detailed requirements apply to all national administrations involved", as well as to private sector companies (Knill and Lenschow, 2000: 3). Even among mainstream scholars the conventional model of environmental statehood was criticised

> for being too intrusive, slowing growth, and delivering more costs than benefits. These critics pointed to the unwanted by-products of regulation, especially its economic impacts. They called for more economic and risk analysis, better ways of setting, and more selective regulatory interventions in the economy.
>
> (Fiorino, 2006: 7)

To summarise, the introduction, after the Second World War, of a conventional model of environmental statehood and regulation took place in a particular phase of global capitalism (i.e. Keynesian forms of welfare and welfare-developmentalism) and in relation to specific socioecological pressures and political demands (i.e. the balance between pro- and anti-commons policies). The state continued to demonstrate an ambivalent stance over the uses of nature, being, as it was, in charge of both environmental protection and the promotion of nature exploitation. This dilemma is described by Offe (1984) as the conflict between the allocative and the productive responsibilities of the welfare state (i.e. a tension between the accumulation and legitimisation functions of the state, which sometimes conflict and affect the interests of capital). In the environmental sector, such tension was translated into the need to extend the expansionist drivers of capitalism and, at the same time, contain the socionatural contradictions of the capitalist economy. After several decades of implementation and contested results, it became clear that the conventional regime of environmental statehood achieved some important outcomes in terms of formalising a system of national environmental control, but also failed to effectively curb the trend of growing socioecological impacts and grassroots reactions. The weakening of conventional environmental statehood coincided with the multiple politico-economic crisis of the capitalist state in the 1970s and 1980s and led to a transition to more flexible approaches. In the next chapter we will explore the conceptual, political and practical elements of this contingent transformation. Before that, it is necessary to analyse the politico-philosophical ideas that underpin conventional environmental statehood in more detail.

The Hobbesian influence

When examined in detail, it becomes clear that the ideological and institutional basis of conventional environmental statehood was neither new, nor different from the political reasoning employed to justify and strengthen the emerging capitalist state centuries before. On the contrary, the main inspiration behind the introduction of a comprehensive and centralised model of environmental statehood was the political theories put forward by Thomas Hobbes, in particular his advocacy of a public authority strong enough to bring rationality and maintain peace between nations. Although the post-war period was characterised by social democratic regimes (at least in the Western world), the environmental statehood adopted by the welfare state paradoxically invoked the Hobbesian formulation of rule and power. The reason for a tacit invocation of the political thinking of Hobbes was the mounting challenge faced by the state in the post-Second World War decades as it struggled to reconcile the urgency of environmental protection with the anti-commons demands of the capitalist socioeconomy. The evident political tension between those two goals required a higher dose of state interference that had to go beyond the solution adopted in the early phase of environmental statehood. The Hobbesian inspiration was less apparent in the first half of the twentieth century because the levels of socioecological disruption were not so pronounced and environmental issues were lower on the political agenda. But with rising rates of environmental problems, and stronger public opinion, the state was compelled to respond in the form of more systematic legislation and enforcement rules. Conventional environmental statehood was essentially a concerted attempt, following the logic of self-preservation of the established order, to contain the increasing degradation caused by production activities, infrastructure expansion and the use of natural resources.

It may sound paradoxical that democratic states in the Western countries had to resort to such strict political philosophy to find answers to environmental problems. Nonetheless, the assessments of environmental degradation by policymakers in this period closely replicated the perplexity of Hobbes with the stupidity of unchecked humans and his defence of the advantages that could be obtained from the interventions of a severe state. For Hobbes (1996), the natural circumstances of people constitute a state of war – considered the *state of nature* still found among the savages in America, who have no government at all and live in a brutish manner – this state being maintained by the absence of a centralised power able to secure religion, properties and morals. The 'state of nature' is the realm of those described by Hobbes as the ignorant and uncontrolled people, who need to be subjected to the civilising and disciplinary role of the state. As famously preached by Hobbes, "the condition of man is a condition of war of every one against every one" (p. 86). Against these deplorable circumstances, reason should prevail through a series of laws, starting with the commandment that a man is forbidden to do that which is "destructive of his life, or taketh away the means of preserving the same; and to omit, that, by which he thinketh it may be best preserved" (p. 86). Consistent with his interpretation of the genesis of violence, Hobbes advocates the departure from the original 'state of nature' towards an anthropocentric world under the control of a centralised state authority. The state is described by Hobbes as the entity needed to settle the brutality of autonomous people and overcome the vicious condition of a society suffering from its proximity to nature (in this case, nature is more than a material world set apart from society, it is a relationship between egoistic individuals and their biophysical condition).

Hobbes described such human status as one of inevitable and fratricidal competition (i.e. a war of every man against every man) where people are found with an intransigently individualistic behaviour and are virtually unable to understand and compromise. For Hobbes (1996: 82), it is a shame that "Nature hath made man so equal" and as a result "men have no pleasure ... in keeping company, where there is no power able to over-awe them all. For every man looketh that his companion should value him, at the same rate he sets upon himself" (p. 83). People are incorrigible beasts that need a strong power ("one man or one assembly of men") to look after them and to make sure that a social contract (the 'convenant') is observed. Due to the failure of humans to negotiate their differences among themselves, the leadership and command of the all-powerful Leviathan – a combination of god, animal and machine that personifies the authoritative state – is both desirable and unavoidable. The state is portrayed by Hobbes as a powerful creature, located above ordinary moral disputes, which operates beyond popular criticism. Therefore, in a situation of growing socioecological impacts, as experienced in the post-Second World War decades, the authority and morality of the state were invoked as the necessary basis for the control of the 'state of nature'. The Hobbesian political model served as a decisive reference for the introduction and consolidation of conventional environmental statehood especially because of the need to have prompt and robust responses to environmental degradation and to secure the corresponding enforcement of regulation. The solution envisaged by Hobbes for the organisation of society and the containment of anarchy served to guide the new regulatory apparatus under implementation since the 1960s (for example, the distinctive Hobbesian character of environmental impact assessment and pollution control introduced in this period). These regulatory instruments were part of the institutional framework needed to overcome the widespread ignorance and brutality ultimately responsible for mounting socioecological degradation.

The ingenious rationalisation of state power, which Hobbes eloquently expanded through elements of religious and regal argumentation, has had a lasting impact on the formulation of the conventional version of environmental statehood and regulation. The environmental agenda of the Keynesian state followed centralised norms and guidelines needed to maintain social and economic activities. A Leviathan-like control by the state was seen as a necessary force to moderate excesses and to reduce sociopolitical reactions, even at the price of contradicting other more liberal policies of the same state. Following Hobbes (1998: 78), those aims could not be achieved merely by common agreement alone, because of the evident wickedness of the human character; hence the need for a strong sovereign authority and consequent penalties (the "Right, The Sword of Justice."). The "security of individuals, and consequently the common peace, *necessarily* require that the *right* of using the *sword* to punish be transferred to some man or some assembly; that man or that assembly therefore is necessarily understood to hold *sovereign power*". Conventional environmental statehood was primarily informed by the constitutional pragmatism of Hobbes and not by the bureaucratism of Max Weber (which was wrongly argued by authors such as Fiorino, 2006). Instead of the Weberian search for efficient procedures, the ultimate goal of the introduction of environmental statehood was the tenacious management of environmental degradation while preserving the anti-commons trajectory of capitalist production. The influence of Hobbes on the consolidation of environmental statehood was not only more relevant than Weber's, but was more appropriate than the equivalent political elaboration of Machiavelli, and also the liberalism of Locke.

At face value, conventional environmental statehood could have been, perhaps, more easily influenced by elements of Machiavellian state theory and the rule of authority. Even the current uses of the word 'state' were firstly adopted by Machiavelli in his description of different nation government configurations (Bobbio, 1995). Following the Platonic advocacy of a powerful authority, Machiavellianism was the rediscovery of the foundations of experience and the excuse for a strong, sometimes violent, means for a supreme end (Arendt, 2006). The reasons behind human actions are the need for security, greed, honour and personal freedom, which are only possible through the unrestricted exercise of power. Rather than a simplistic apology for authoritarian force, Machiavelli revealed "the necessity and the autonomy of politics that is beyond good and moral evil, that has its laws against which is futile to rebel" (Croce, 1925, in Mariátegui, 2011: 200). Machiavelli was one of the early architects of the idea of a European state system, which was intended to be primarily effective and resilient, and only then, after those criteria were met, needed to occupy itself with justice and the desires of the general public. Machiavelli's pragmatism is based on the separation between moral goals and concrete action, which in practice means that the state should pursue a righteous cause, but when law is not sufficient, it is perfectly justifiable to resort to force ("a prince must know how to make use of the nature of the beast", Machiavelli, 2005: 60).

However, there is an important difference between Hobbes and Machiavelli, given that the former aimed to contain the brutality and disorganisation of people, whereas the latter wished to split and control the people themselves. The state, personified in the figure of the Machiavellian prince – the prince as the despotic, but efficient, ruling system, rather than only one single ruler – is not only entitled, but expected to make use of violent measures in order to contain inconvenient oppositions. The practical consequence is that an environmental statehood inspired by Machiavelli would become not only onerous and unpopular, but could trigger a counterproductive reaction that in the end may lead to the disregard of environmental regulation and the inability of the state to navigate its tenuous pro- and anti-commons responsibilities. Should environmental regulation follow Machiavellian ideas, the control of natural resources and recovery of ecological conditions would be commanded from the centre of the state apparatus in a highly generic, nonspecific and unproductive way. Machiavelli recommended that the prince should also avoid being despised and hated by the people, but the insistence on the supra-moral authority of the state would lead to environmental policies largely cut off from the actual condition of local groups and socionatural systems. The Machiavellian state is the ultimate bastion of force and, if necessary, it must defeat both political opposition and remove barriers that prevent access to ecosystems and ecological processes.

Anyone who becomes a master of a city accustomed to living in liberty and does not destroy it may expect to be destroyed by it, because such a city always has a refuge in any rebellion in the name of liberty and its ancient institutions.

(Machiavelli, 2005: 19)

According to Machiavellianism, public policies must fulfil the political, territorial and operational needs of the state, which is only vaguely required to be connected to the necessities and the expectations of those under its rule. Any unreasonable reaction against state objectives should be decisively and appropriately suppressed. For Machiavelli, once a territory or ruling position is secured, there are three main ways of holding it and maintaining state functioning: through the bare destruction of the existing situation, through a more direct contact and involvement with the object of control, or a compromise between limited freedom, law enforcement and tax collection. As a result, while Machiavelli incorporates evilness into the very operation of the state, Hobbes sets the state against the incidence of evil (i.e. evil is external, distinct from the state). An environmental statehood directly informed by Machiavellism would primarily focus on the containment of the irrational behaviour and political resistance of those dissatisfied with the state, rather than on the organisation of a centralised Hobbesian-informed authority capable of both firm control and political mediation. If Machiavelli reasoned about the necessity of a pragmatic authority, Hobbes justified the strong government needed to control the chaotic and brutish condition of people still in a 'state of nature'. With Hobbes, the political philosophy of the strong state previously anticipated by Machiavelli was further developed and went beyond territorial conquests in favour of unwavering governments capable of dealing with brutality and ignorance.

In the same way, Locke's unreserved defence of private property and the right to rebel against an excessively intrusive state did not easily fit with the needs of an effective environmental statehood. On the one hand, the influence of Locke played a main role in state policies and legal frameworks related to the anti-commons project that permeates capitalist economy. On the other hand, Lockean economic liberties could easily undermine the enforcement of environmental protection rules required for the long-term viability of capitalist society. For Locke, private property was achieved through hard work and the conversion of earth resources into useful goods. Social inequalities are the resulting condition of political interaction and the amount of labour invested in private properties. Private property and the related level of personal wealth cannot be constrained by the state, which should only effect socioeconomic changes by consent. In that regard, the main role of the state is to guarantee the opportunity that all citizens have to acquire and sustain private property. People are equal by nature and, in contrast to Hobbes's focus on scarcity and conflicts, the Lockean state of nature is a situation of high levels of morality and reason. In the famous Second Treatise, Locke (1980) argues that men are free by nature and enjoy the rights to life, liberty and property. Governments exist by command of the people in order to protect personal rights and promote the public good, and, therefore, these goals cannot be the antithesis of the benefits derived from the interaction of free people in their pursuit of property and wealth. It is precisely this emphasis on a limited role for the state apparatus, amid major economic inequalities, that meant that Hobbes's influence on the organisation of conventional environmental statehood was greater than Locke's.

There was actually the need in the post-Second World War period for a (formally) strong, centralised environmental authority to set limits on private property and also to maintain a functional and productive connection between the ruler and those ruled. It was in that context that Hobbes became the main influence on conventional environmental statehood, given that in his view the state is a powerful organism responsible for overseeing collective well-being and securing social order beyond the 'state of nature'. It is implicit in the Hobbesian philosophical construction that people must accept the negative character of their behaviour and consent to lay down their rights in favour of the wisdom and authority of the state. Following Hobbes, environmental problems can never be properly understood by the general public, which justifies the overwhelming intervention of the 'big brother' Leviathan on the behalf of all 'savage' citizens. Hobbes (1996: 99) submits that "Whatsoever is done to a man, conformable to his own will signified to the doer, is no injury to him". Whereas Machiavelli educated a prince to behave like a prince and, when necessary, do all the injuries at once (such as allowing the construction of new nuclear power plants and roads through conservation areas, under the need to keep the economy growing), Hobbes provided the necessary consolation for the simple citizen to assent to powerful impositions coming from the political centres (such as additional taxes on water, energy and waste to fund obscure and potentially untrustworthy branches of the state apparatus).

The contribution of Hobbes to the introduction of conventional environmental statehood did not mean a mechanical acceptance of a repressive, autocratic form of regulation, but rather a more sophisticated centralisation of authority and decision-making in the hands of the state. Conventional environmental statehood informed by Hobbes served to demonstrate the power and leadership exerted by the state to inhibit environmental degradation, although it also revealed the shortcomings of centralised approaches for dealing with the mounting socioecological problems. Making reference again to the NEPA legislation, it has been considered the 'Magna Carta' of USA's environmental law because of its provision that federal agencies should assess the environmental consequences of their proposed actions. Before NEPA, there was only limited legislation protecting national parks, national forests and wildlife in the United States. The new Act occupied a whole decade of congressional debate – the 1960s – and had major repercussions for the regulation of air and water pollution, endangered species and resource conservation. NEPA effectively changed the landscape of environmental politics and, interestingly, rebalanced the equilibrium of power between capitalists and environmentalists, albeit within the perennial tensions between pro- and anti-commons targets. In effect, after its introduction, there was a significant increase in terms of litigation, particularly that initiated by environmental groups aiming to halt or delay projects considered unsound. The centralised framework did not mean the new regulation was easy to interpret and there were many cases where substandard, opaque, hard-to-understand NEPA document writing has resulted in controversy and contestation (Hansen and Wolff, 2011).

The NEPA example makes it evident that Hobbesian absolutist liberalism subtly provided the basis for the introduction of environmental legislation and the consolidation of a dedicated regulatory apparatus by the state. The justification of the firm hand of the state and its defence of social stratification as a productive social force had significant repercussions for environmental regulation as an attempt to restrain individual liberties in order to guarantee social and economic survival. Hobbes specifically claimed that people should delegate their autonomy to the sovereign power in the search of a 'superior' social order. This act of 'delegation' was an apt metaphor for the supposed, ideological advantages of centralised environmental regulation and rigid socioecological strategies, even if this meant eroded socioeconomic and political rights. The state could then become the guardian of society and the caretaker of the environment in order to overcome the 'state of nature'. The state was expected to contain the destructive drive of economic production and protect the material, ecological conditions under which the economic appropriation of nature had to proceed. It is clear that such Hobbes-inspired regulation was introduced at the expense of the autonomy and creativity of the individual. In that process, "even though the individual disappears before the apparatus which he serves, that apparatus provides for him as never before" (Adorno and Horkheimer, 1997: xiv).

Although Hobbesian philosophy furnished the state with a centralised state-fix in the form of codes, fines and penalties, it eventually proved unable to deal with growing rates of impacts and mounting environmental risks. The main limitation of Hobbesian reasoning was the subordination of the presumed brutality of people (i.e. the state of wildness experienced by ungoverned men) to the greater brutality of the state, which is thenceforth allowed to contain human folly with the full weight of its power. The responses formulated to environmental disruption followed the prerogative of the state to identify or ignore environmental problems according to its broader economic goals and political commitments. The Hobbesian defence of a strong, overpowering state that is needed to remove disorder and mismanagement of natural resources has the side effect of directly inserting nature into the realm of economic fetishism and social control. Some key shortcomings of the NEPA regime, for example, were the fact that procedural requirements overtook the more substantive environmental goals, the difficulty to engage the public in a meaningful assessment of alternatives, and the persistent interference of the stronger interest groups trying to contain the impact of environmental regulation.

> Sometimes, NEPA-generated scientific findings are revised or suppressed by political pressures.... Issues like global warming, sustainability, endangered

species, transportation, energy policy, and population growth can be influenced as much or more by political ideology as by scientific findings.

(Hansen and Wolff, 2011)

The complex origins of the conventional environmental statehood in the post-Second World War years explain the partial containment of environmental impacts and the surfacing of new forms of social unrest. In the end, what was necessary was a transition to more responsive forms of environmental statehood and regulation, which based itself on the political ideas of Kant and the sophisticated, slightly contradictory, combination of strong morality and a gradual approximation to the ideal state apparatus, as discussed in the next chapter. But first let's consider the achievements and failures of a Hobbes-informed experience of environmental statehood introduced in the southeast of Brazil.

Environmental statehood and urban development in the Baixada Fluminense (Brazil)

This section illustrates how Hobbesian forms of social contract influenced the introduction of environmental statehood even beyond the Western world. It is also intended to demonstrate that environmental statehood is more than simply the control of harmful activities, and actually encompasses the intricate trialectics between state, society and the rest of nature. In addition, the environmental statehood put in place in the Baixada Fluminense – a densely populated wetland area (with more than three million residents) to the northwest of the city of Rio de Janeiro – vividly reveals the entirety of socionatural relations mediated by the state according to particular historico-geographical situations. The Baixada Fluminense is formed by eight municipalities located to the western side of the Guanabara Bay and has parts of its terrain below sea level, as well as a large extension of its watercourses affected by the tidal regime. In the early days of the Portuguese colonisation, plantation farms were established in the Baixada – after the displacement in the 1550s of the indigenous Tupinambá tribes – for the production of sugarcane. The violence towards people was followed by violence towards land and water in the form of deforestation, drainage of swamps and the opening of roads. The primitive state and the first generations of migrants acted together in the appropriation and transformation of water bodies according to the overall colonisation goals. In the eighteenth century, the local river system was intensively used as a corridor for the transport of gold, and then coffee, from the inland to the Guanabara Bay and then to the port of Rio de Janeiro.

For a long time, the insalubrious conditions of the Baixada were responsible for low population densities and the prevalence of rural activities. This started to change in the second half of the nineteenth century with public interventions in terms of land reclamation, the construction of railway lines and the launch of early sanitation agencies. The need to improve transport connections with the city of Rio encouraged the creation, since 1854, of the first Brazilian railway along the lowlands of the Baixada Fluminense, followed by additional rail lines in the next decades. Around the turn of the twentieth century, the construction of bridges and river passages through swamps and watercourses had further reduced the prospects of in-stream navigation. With the decline of navigation, the provincial government began the drainage of parts of the wetland to reduce the incidence of waterborne diseases, malaria in particular, associated with the water which had accumulated along the railways (Fadel, 2009). In 1910, a technical commission was established to plan the recovery of the river system and propose ways to stimulate agriculture production and commercialisation using the river network. This had parallels with the introduction of an early model of environmental statehood in the Clyde River Basin, but happened a few decades later than the Scottish experience (see above). Between 1910 and 1916, a German company was specifically contracted to dredge, clean and interconnect the local rivers (the company had to interrupt its operation due to Brazilian alliances during the First World War). Because of the heavy costs involved and changes in the federal administration, large parts of the planned government interventions were abandoned and only isolated restoration works were effectively carried out.

State interventions in the Baixada evolved from land reclamation for farming and river navigation (in the nineteenth century) to river engineering, flood defence and urban water supply (in the twentieth century). After the abolishment of slavery in 1888, the area was one of the preferential sites where ex-slaves and migrants from other parts of the country could settle their families and find a place to live. Land reclaimed in the Baixada offered the cheapest alternative to a significant proportion of poor migrants that arrived with little more than their own labour power. With the national political insurrection of 1930, the previous macroeconomic model based on the export of coffee was gradually replaced by an emphasis on 'import substitution'. This required changes also to the allocation and management of territorial resources. In fact, during the industrialisation period, the national state apparatus played a fundamental role in promoting economic growth, especially in the southeast region of Brazil. A system of polders and dykes had been built in the 1930s to facilitate food production for the metropolitan region of Rio de Janeiro, but it was gradually dismantled and engulfed by the chaotic urbanisation along riparian areas and floodplains. At the same time, with the construction of the modern highway connection between São Paulo and Rio de Janeiro in 1951 – which crosses the Baixada – new areas became available to accommodate the relentless flow of incoming migrants.

Immigration peaked in the 1950s and 1960s, during which time population growth reached rates as high as 10 per cent per annum. The majority of the low-income residents built their houses themselves over a long period of time and expanded whenever there was some disposable money to spend. With high rates of urbanisation in the second half of the twentieth century, the chronic problems of flooding and insalubriousness became deeply coupled with human-caused water scarcity (i.e. lack of reliable potable water). Deficient water infrastructure persisted during the whole twentieth century and became even more evident after the inauguration of a large oil refinery in the municipality of Duque de Caxias in 1961, which required the construction of two dedicated adduction pipelines to secure water for its own operation and that of associated industries. Water degradation and insufficient water services were the final outcome of a long process of appropriation and transformation of the regional river network in order to produce food, facilitate navigation and reclaim land for urban expansion. The complex and multifaceted socioecological problems of the Baixada are intensely linked to the insufficiencies and constraints of state action. Government action was notoriously partial, selective and even discriminatory, with most of the investments and responses from public agencies serving the demands of stronger groups and locations. In order to achieve those objectives, a Hobbesian-informed approach was necessary.

The failure to protect the local river catchments and to provide safe water and sanitation to the population serves as a revealing entry point into the limitations of the centralised, conventional environmental statehood introduced in the post-Second World War period. A series of regulatory measures and investment programmes were directed towards environmental management issues in the Baixada, which mobilised large sums of capital and created great expectations among the residents. However, those interventions were largely frustrated by a combination of bad administration, demagogic appropriation of public policies and the persistent neglect of grassroots demands (Ioris, 2011). A perverse combination of disorganised urban growth in a tropical wetland with a precarious infrastructure resulted in favourable conditions for the proliferation of serious and recurrent socioecological problems. The end result was a highly populated area that suffers from consistent problems of flooding (in the summer) and water deficit (in the winter, the dry season), whilst a large extension of the river system was seriously affected by organic and industrial pollution (with severe levels of chemical pollution, high levels of faecal coliform, low levels of oxygen in the water and contamination of sediments by heavy metals). Because of mounting river degradation and high population inflow, the Baixada shifted from being a water exporter to Rio de Janeiro in the nineteenth century to a net importer of 90 per cent of its own water supply at the end of the twentieth century (most water now comes from the Guandu River, which in turn depends on the transference of water from the Paraíba do Sul River Basin). Those not officially served by mains water had to rely on a combination of boreholes, water sellers, unauthorised connections to the public network and various forms of joint action between neighbours.

The problems of urban growth and environmental management in the Rio de Janeiro metropolis reflected a pattern that is still common to all the large cities of Brazil, where social inequalities are closely related to the priorities of market transactions, and systematic discrimination. In this specific case, the difficult socioecological situation of the Baixada Fluminense was the consequence of ill-conceived development policies and the unresponsiveness or inappropriateness of strategies developed along the lines of conventional environmental statehood and regulation. Government agencies seemed beleaguered with the complexity of social and environmental problems and had major difficulty in taking into account the dissatisfaction of the local residents, who resentfully complained about the discrimination they experienced. Such confinement of large percentages of the population in marginalised, unsuitable zones can be directly related to what Agamben (1998) defines as the "spatialisation of the political exception". To fully understand this, it is necessary to move back in time to demonstrate how the interconnections between state, society and the rest of nature in the Baixada have been influenced by the narrow, top-down basis of conventional statehood. During the main immigration period, populist approaches had already proved to be very effective with a contingent of illiterate, impoverished and disorganised migrants. State interventions were restricted to some isolated investments in public infrastructure, a trend that continued during the military dictatorship that controlled the country between 1964 and 1985.

In the 1970s, public works in the Baixada represented a significant proportion of the total amount of resources allocated to water infrastructure in the State of Rio de Janeiro, but the distribution of funds and the operation of water services primarily favoured the wealthier, and politically stronger, places (Marques, 1996). Government initiatives were directed towards the locations already served by some form of public infrastructure, primarily under the justification that the state could not make investments where the legal status of property was uncertain. Given the limited political activity during the dictatorship period (e.g. the state governor and many local mayors of the Baixada were directly appointed by the central government), the majority of the local population could neither convey their demands, nor protest about the lack of water supply and sanitation. The only residual activity existed in the realm of the Catholic Church and under the protection of the progressive bishops of the cities of Duque de Caxias and Nova Iguaçu. The final years of the dictatorship (after 1979) were a phase of intense political revival and reaction to the democratic deficit: hundreds of neighbourhood associations and, shortly after, the respective municipal associations were created and began to play an important role in bringing popular demands to the attention of the authorities responsible.

During the 1980s, popular mobilisation (primarily expressed through territorialised action around neighbourhood associations) was able to exert some influence on the direction and rate of investments, particularly in the public water utility (CEDAE). However, this period of higher responsiveness from public agencies was very short-lived: a combination of turbulent transitions from one administration to another and a lack of genuine managerial commitment resulted in a discontinuity of projects, notorious corruption and a huge waste of resources. Since then, most of the interventions of governmental agencies responsible for water services and environmental protection have amounted to a series of piecemeal, top-down responses. For instance, the first governor elected at the end of the dictatorship (Leonel Brizola) seized the opportunity to consolidate his authority in the Baixada and was quick to recognise the strength of the newly created neighbourhood associations and of the Political Committee of Sanitation of the Baixada Fluminense (a grassroots organisation dedicated to debating water and sanitation issues). By offering paid jobs in the government to the most prominent leaders, those organisations were soon under the control of the state government and had their autonomy gradually undermined (Porto, 2003).

In 1984, the Global [i.e. Comprehensive] Sanitation Plan (PEB) was launched by Governor Brizola with ambitious targets and an innovative focus on condominial sanitation schemes (i.e. a low-cost technology through which pipelines are laid under paths and across properties, rather than under roads, using local labour). However, clashes between state and federal administrations destabilised the implementation of the PEB and, out of the 576 km of pipelines initially planned, only 70 km were effectively installed. The next administration (Governor Franco) redirected the PEB project to a more conventional technological design and reduced the overall target to 251 km (which was also not achieved). The problems of bad sanitation services, limited involvement of the population and a problematic relationship between state and municipal administrations nonetheless persisted. CEDAE continued to prioritise investments in the wealthier locations within the metropolitan area, which obviously did nothing to help the Baixada. A new initiative (called Sector Water Supply Plan) was launched by Franco and aimed, but failed, to install 89 km of pipelines. In addition, in 1988, Governor Franco introduced another project, Reconstruction Rio, in response to the outcry over the floods that castigated the Baixada two years earlier, with a budget of US$288 million (mostly funded by the World Bank) for sanitation, urban drainage and solid waste. But only around half of that amount was effectively spent due to bureaucratic delays and, more importantly, the end of the governor's term of office. Brizola returned as governor in 1990, but his second mandate was again marred by a tumultuous relationship with the federal government.

Despite the repeated problems that dogged the various projects formulated in the 1980s and 1990s (note that only a subset is mentioned here), a much larger scheme was launched in 1994, the Guanabara Pollution Control Programme (PDBG) with a total budget of US$860.5 million. The focus of the new Programme was the whole river system that drains to the Guanabara Bay and, specifically for the Baixada, PDBG included major infrastructure works, such as seven new storage reservoirs (to serve a population of 575,000), two sewage treatment works (to serve a population of equivalent size), drainage, planning, environmental restoration and educational projects (Rio de Janeiro, 1994). Unfortunately, five administrative mandates later (these were State Governors Brizola, Alencar, Garotinho, Rosinha and Cabral), PDBG is still not concluded, having been hindered by constant delays, lack of dialogue with civil society, failure to engage the local authorities and ill-conceived plans that are not easily connected to the existing water infrastructure (Britto, 2003). Projects have been fraught with serious evidence of corruption, waste of resources and the overlapping of targets from one project to the next (Vargas, 2001). In addition, local authorities have systematically failed to apportion resources to complement investments made by the state. These distortions in the programmes of investment have been aggravated by the mismanagement of the water utility and its inability to adapt to the demands of the impoverished communities of the Baixada. CEDAE is well known for its insensitivity to social criticism and permeability to the interests of private construction companies.

Despite the fact that between the 1980s and the 1990s public investments totalled around US$1.5 billion – mostly funded by international loans from the World Bank and the Inter-American Development Bank – those interventions contributed little to resolve the problems of pollution, scarcity and flooding. Provincial and federal governments put emphasis on the physical expansion of the water infrastructure, but engineering works were planned in isolation from city planning and suffered from systematic interruptions and evidence of corruption. As a result, a significant proportion of the residents still had to resort to alternative sources of water, such as purchasing it from water vendors or drilling boreholes. The systematic announcement of new projects for the same problems and the same locations – without resolving the structural deficiencies and without ever changing the stratified, authoritarian relationship between state agencies and local communities – served the double purpose of containing popular criticism and creating a permanent source of political profit. The perpetuation of precarious environmental conditions has transformed popular demands into an enduring, and profitable, political machinery that operates intermittently both during electoral campaigns (e.g. with promises of new investments) and between elections (e.g. occasional and paternalistic provision of water lorries by politicians in exchange for votes in the next election, which are either paid for with public money or provided by the water utility according to political influence). The more productive dialogue between the government and the neighbourhood associations that was established during the early stages of redemocratisation in the 1980s rapidly declined and was replaced by circumstantial and ephemeral forms of communication. Recent interventions generally ignore any lessons learned in earlier projects, preferring to follow a short, bureaucratised appraisal of previous experiences.

The experience of the Baixada Fluminense provides an example of the complex evolution of a Hobbesian-inspired environmental statehood and its failure to meet the socionatural requirements of the wider contingents of low-income social groups (particularly the restoration of the watercourses, reduction of pollution, management of flood risk and supply of safe drinking water). Through the interventions of a strong, authoritative state the old traditions of centralisation and paternalism of Brazilian politics have been revived and applied to the mediation of environmental issues. The evident failures of government initiatives – including the federal, the state (provincial) and the municipal levels of government – were more than operational incompetence, but constitute a coherent feature of the relation between state and society articulated through conventional environmental statehood and regulation. The state, in the person of elected or unelected governors, promised to resolve the backlog of socioecological problems, but environmental conservation and public services were persistently subordinate to stronger economic and political interests. The ambiguities of environmental statehood in the Baixada, which unfolded through the capricious mixture of neglect and populism that characterised the welfare-developmentalist policies in the southeast of Brazil during most of the twentieth century, were further nourished by the already weak forms of public mobilisation and the continual attempt to domesticate grassroots leaders and isolate the more critical voices. The same state, which represents primarily the interests of the conservative elite living outside the Baixada, is also guilty of stimulating a particular form of regional development based on low salaries, fiscal incentives and an abundant workforce. In that sense, the management of natural resources in the Baixada is an integral element of the public policies and class-based struggles mediated by the capitalist state in charge of the industrialisation and the economic development, which makes only piecemeal concessions to popular demands.

The example of the Baixada Fluminense also demonstrates the need for an explanatory tool that comprehensively articulates the tensions between social dynamics, state authority and environmental change, that is, the trialectics between state, society and the rest of socionature mentioned in the previous chapter. During the implementation of a conventional model of environmental statehood, local environmental problems remained part of a potentially explosive, but in practice feeble, combination of authoritarian anti-commons rule with the reclamation of critical commons (i.e. water, air and riparian land). Following the Hobbesian influence, state interventions were always stratified, beyond contestation and effected through regular investments in infrastructure and erratic restoration plans. Such initiatives were never debated upfront with local communities, but were based instead on an expert interpretation of problems and a politicised prioritisation of targets. The chaotic 'state of nature' of the Baixada Fluminense (to repeat the most evocative Hobbesian term) was actually created by the welfare-developmentalist state in its effort to tame rivers and catchments as well as to organise existing and newly arrived social groups. The multiple problems of this state of nature produced by the state required concerted responses from the same state apparatus. However, the conventional configuration of environmental statehood achieved only isolated results, whilst at the same time perpetuating socioecological tensions manifested from household and municipal level to a national and international extent. The apparent shortcomings of conventional statehood triggered a transition to a new phase of the anti-commons agenda of the capitalist state, as will be examined in the next chapter.

Note

1 In historical terms, it is interesting to observe that there was a gap between the interventionist rules on economic and social matters (coinciding with the attempts to restore economic order after the collapse of classical liberalism in the post-First World War phase) and the introduction of more strict environmental statehood in the post-Second World War period. The main reason for this time-lag was the fact that environmental disruption had actually been accelerated by economic growth policies promoted by the Keynesian state, which included heavy infrastructure works (dams, roads, ports, etc.), agroindustrial expansion and intensification, and higher circulation of commodities. This means that the environmental statehood introduced by the, then, mature Keynesian state was also a response to the ecological excesses of the early phase of the Keynesian type of economic stimulus. It demonstrates that the fundamental aim of environmental statehood at this time was to deal with the most immediate threats to the existing socioeconomic relations and to contain rising political criticism.

Bibliography

Adorno, T.W. and Horkheimer, M. 1997. *Dialectic of Enlightenment.* Trans. J. Cumming. Verso: London and New York.

Agamben, G. 1998. *Homo sacer: Sovereign Power and Bare Life.* Trans. D. Heller-Roazen. Stanford University Press: Sanford.

Arendt, H. 2006. *Between Past and Future.* Penguin: New York.

Arrighi, G. 1994. *The Long Twentieth Century: Money, Power, and the Origins of Our Times.* Verso: London and New York.

Asher, K. and Ojeda, D. 2009. Producing Nature and Making the State: Ordenamiento Territorial in the Pacific Lowlands of Colombia. *Geoforum,* 40(3), 292–302.

Bobbio, N. 1995. *Stato, Governo, Società: Frammenti di un Dizionario Politico.* Einaudi: Torino.

Britto, A.L.P. 2003. Implantação de Infra-estrutura de Saneamento na Região Metropolitana do Rio de Janeiro. *Estudos Urbanos e Regionais,* 5(1), 63–77.

Callinicos, A. 2007. Does Capitalism Need the State System? *Cambridge Review of International Affairs,* 20(4), 533–49.

Fadel, S. 2009. *Meio Ambiente, Saneamento e Engenharia no Império e na Primeira República.* Garamond: Rio de Janeiro.

Fiorino, D.J. 2006. *New Environmental Regulation.* MIT Press: Cambridge, MA.

Foster, J.B. 2002. *Ecology against Capitalism.* Monthly Review Press: New York.

Giannotti, J.A. 2011. *Marx Além do Marxism.* L&PM: Porto Alegre.

Gramsci, A. 1971. *Selections from the Prison Notebooks.* Trans. Q. Hoare and G.N. Smith. Laurence and Wishart: London.

Hansen, R.P. and Wolff, T.A. 2011. Reviewing NEPA's Past: Improving NEPA's Future. *Environmental Practice,* 13(3), 235–49.

Hobbes, T. 1996 [1651]. *Leviathan.* Oxford University Press: Oxford.

Hobbes, T. 1998 [1642]. *On the Citizen.* Ed. and trans. R. Tuck and M. Silverstone. Cambridge University Press: Cambridge.

Ioris, A.A.R. 2001. Water Resources Development in the São Francisco River Basin (Brazil): Conflicts and Management Perspectives. *Water International,* 26(1), 24–39.

Ioris, A.A.R. 2011. Values, Meanings and Positionalities: The Controversial Valuation of Water in Rio de Janeiro. *Environment and Planning C,* 29(5), 872–88.

Ioris, A.A.R., Hunter, C. and Walker, S. 2006. A Framework of Indicators to Assess the Sustainability of Freshwater Systems. *Physical Geography,* 27(5), 396–410.

Knill, C. and Lenschow, A. 2000. *Implementing EU Environmental Policy: New Directions and Old Problems.* Manchester University Press: Manchester and New York.

Krenke, A.N. and Chernavskaya, M.M. 2010. Environmental Legislation in Russia during the Period 1855–1913. *Izvestiya Akademii Nauk, Seriya Geograficheskaya,* 5, 17–30.

Leiss, W. 1994. *The Domination of Nature.* McGill-Queen's University Press: Montreal.

Locke, J. 1980 [1690] *Second Treatise of Government.* Hackett: Indianapolis.

Machiavelli, N. 2005 [1513]. *The Prince.* Trans. P. Bondanella. Oxford University Press: Oxford.

Manheim, F.T. 2009. *The Conflict over Environmental Regulation in the United States: Origins, Outcomes, and Comparisons with the EU and Other Regions.* Springer: New York.

Mariátegui, J.C. 2011. *An Anthology.* Ed. and trans. H.E. Vanden and M. Becker. Monthly Review Press: New York.

Marques, E.C. 1996. Equipamentos de Saneamento e Desigualdades no Espaço Metropolitano do Rio de Janeiro. *Caderno de Saúde Pública, Rio de Janeiro*, 12(2), 181–93.

Marx, K. 1976 [1867]. *Capital: A Critique of Political Economy*. Vol. 1. Trans. B. Fowkes. Penguin: London.

Meadowcroft, J. 2005. From Welfare State to Ecostate. In: *The State and the Global Ecological Crisis*, Barry, J. and Eckersley, R. (eds). MIT Press: Cambridge, pp. 3–23.

Offe, C. 1984. *Contradictions of the Welfare State*. Hutchinson: London.

Pádua, J.A. 2002. *Um Sopro de Destruição: Pensamento Político e Crítica Ambiental no Brasil Escravista (1786–1888)*. Jorge Zahar: Rio de Janeiro.

Panayotakis, C. 2011. *Remaking Scarcity: From Capitalist Inefficiency to Economic Democracy*. Pluto Press: London.

Poggi, G. 1978. *The Development of the Modern State: A Sociological Introduction*. Stanford University Press: Stanford, CA.

Porto, H.R.L. 2003. *Saneamento e Cidadania: Trajetória e Efeitos das Políticas Públicas de Saneamento na Baixada Fluminense*. FASE: Rio de Janeiro.

Rio de Janeiro. 1994. *Programa de Despoluição da Baía de Guanabara*. Governo do Estado: Rio de Janeiro.

Selznick, P. 2011. *TVA and the Grass Roots: A Study in Politics and Organization*. Quid Pro Books: New Orleans.

Vargas, L.A. 2001. *O Programa de Despoluição da Baía de Guanabara: Uma Análise na Perspectiva de Saúde Coletiva*. Unpublished PhD thesis. UERJ: Rio de Janeiro.

Wells, D. 1981. *Marxism and the Modern State: An Analysis of Fetishism in Capitalist Society*. Harvester Press: Sussex and Humanities Press: New Jersey.

While, A., Jonas, A.E.G. and Gibbs, D. 2010. From Sustainable Development to Carbon Control: Eco-State Restructuring and the Politics of Urban and Regional Development. *Transactions of the Institute of British Geographers*, NS 35, 76–93.

Wurzel, R.K.W. 2008. European Union Environmental Policy and Natura 2000. In: *Legitimacy in European Nature Conservation Policy: Case Studies in Multilevel Governance*, Keulartz, J. and Leistra, G. (eds). Springer, pp. 259–82.

4 Responding to regulation rigidity and persisting socioecological problems

The Kantian basis of the transition

The search for flexible environmental statehood

The previous chapter considered the institutional strategies devised by the state to cope with the growing environmental disruption associated with urbanisation, intensive production and changes in land use over approximately three-quarters of the twentieth century. The state apparatus, especially in the Western world, was compelled to gradually adopt and implement an environmental agenda in order to protect and restore ecosystems and territorial resources. Its responses came in the form of environmental statehood, which was contingent upon the politico-economic circumstances of the time and operationalised as 'state-fixes' (comprising a range of policies, legislation, licences, environmental impact assessments, etc.). As previously defined, the state-fix is the visible, phenomenological expression of the deeper ideological and political elements of environmental statehood. The early model of environmental statehood coincided with the Second Industrial Revolution, which was itself associated with the consolidation of the mass production economy. The state-fix of this phase was still characterised by patchy environmental regulation and disjointed attempts to organise access to resources and safeguard ecological features. A more substantial response to mounting environmental problems came in the form of the environmental statehood adopted after the post-Second World War economic recovery. This was the conventional model of environmental statehood which was intended to complement other social and economic policies promoted by the welfare state in Northern countries and the welfare-developmentalist in the Global South. In this process, larger and more dedicated segments of the state apparatus were put in charge of increasingly complex environmental matters.

In spite of that, the overall aims of environmental statehood continued to be the containment of the excesses of private property and the preservation of the prevailing relations of production and reproduction. Because of the narrow remit of the conventional model of environmental statehood, the persistence and deepening of the environmental issue became central to public uneasiness about many aspects of the mass consumption society. To large sections of the public, the trend towards socioecological degradation was closely associated with the maelstrom of modern life and the disturbing, uncertain transformations of the planet. A conflict between the promotion of development and the partial mitigation of the impacts of development persisted during the implementation of conventional environmental statehood. On the one hand, state-led economic growth led to intensified production and circulation of commodities, which were connected with the provision of public services, high levels of employment and large infrastructure projects carried out by the state. On the other hand, the state was expected to deal with the negative effects and shortcomings of its own policies and also the impacts of private economic enterprises. At the same time as environmental degradation continued to increase and new types of environmental problems were identified (e.g. the ozone hole, climate change, endangered species, etc.), the response from the state was largely restricted to new environmental legislation, enforcement rules and regulatory agencies. For instance, at least fourteen environmental ministries and agencies were launched in industrialised countries between 1970 and 1972, and more than half of the American states, together with Japan and Western Europe, adopted more detailed environmental legislation (Brenton, 1994). Moreover, the intervention of most government agencies achieved only superficial environmental damage mitigation and marginal compensation for ecological disruption. In practice, the national state continued to promote infrastructure construction, the expansion of economic sectors and the creation of public and private enterprises, whilst timidly trying to inhibit the impacts of developmental strategies and production activities.

The expansion of conventional environmental statehood left an evident tension between the interventionism of the state and the unfulfilled demands of private economic sectors. Although the associated state-fixes varied between countries, the answers provided by politicians, bureaucrats and powerful groups of interests around the globe reflected a similar attempt to remediate environmental problems through mainly centralised and punitive strategies. Because of its modest achievements and intrusive regulation, the basis of conventional environmental statehood started to be increasingly questioned by local, national and international economic players. There was a growing realisation that the interventions of the state were neither achieving socioecological goals nor addressing the changing needs of production sectors. Environmental legislation and the work of regulatory agencies were blamed for being too ambiguous, heavy-handed and bureaucratic. The exhaustion of the Keynesian economic policies and the constant failure of rigid environmental regulation only served to aggravate the hitherto unresolved environmental impacts. The changing mood of the times was a result of the diminishing ability of Keynesian policies to deal with challenging socioecological issues, particularly the social unrest associated with environmental degradation, and new economic pressures in a world that was increasingly globalised and interconnected.

By the mid-1980s, it was evident that the limitations of conventional environmental statehood called for a speedy reform in order to cope with old and new environmental problems. The inadequacies of environmental regulation were linked to the spatial disjuncture between national territories and the space taken up by ecological problems, together with persistent exploitation of resources and the apathy of wider society (Paterson *et al.*, 2006). The diminishing legitimacy of welfare and welfare-developmentalist states in environmental matters also coincided with new forms of environmental negotiation and pressures from civil society at national and international levels (Hurrell, 1994). Ultimately, the limited possibilities of conventional environmental statehood were evidence of how difficult it was to reconcile multiple socioecological demands and associated forms of contestation through primarily top-down, centralised approaches. With the apparent shortcomings and fading legitimacy of Hobbesian-style conventional statehood, the always difficult marriage between private and collective agendas in a capitalist society was finally exposed and a rethinking of environmental statehood became pressingly necessary. There was a perceived need to move away from the narrow control of capital–labour relations into a more responsive eco-state formation capable of reworking state–socionature connections (While *et al.*, 2010). Adjustments in the environmental sector coincided with wider changes in the state apparatus and were part of the redraft of the public–private divide of Keynesian policies.

Twenty years after the first global conference on the environment in Stockholm – which at the time represented the pinnacle of conventional environmental statehood thinking – a new, more ambitious summit took place in Rio de Janeiro in 1992. The United Nations Conference on Environment and Development institutionalised sustainable development as the main concept on which to base environmental policy-making and management approaches hereafter. Sustainable development and other correlated concepts were the unmistakable proof of a movement towards a more flexible, liberal model of environmentalism and environmental statehood. In contrast to the prescriptive, centralised responses of the previous decades, the search for sustainability incarnated the emerging argument for fresh new associations between individuals and the state apparatus. From the perspective of those in charge, sustainable development seemed to provide the rationality needed for dealing with environmental issues according to legitimation and innovation criteria. It provided an opportunity for more sophisticated interventions and appropriate environmental leadership that could be related to the promotion of a liberalised economic order and the affirmation of market-based solutions (including, obviously, market solutions to environmental degradation and conservation). The verticalisation of the Hobbesian-informed approaches was replaced with a more horizontal collaboration between state and non-state players, such as business sectors, NGOs, think tanks and so on.

The incorporation of sustainable development into policy-making was primarily achieved through the adoption of the discourse and practice of environmental governance. The European Commission (2001), for instance, defines governance as "rules, processes and behaviour that affect the way in which powers are exercised at European level, particularly as regards openness, participation, accountability, effectiveness and coherence". Governance, instead of the usual government interventions associated with conventional environmental statehood, represented a set of flexible strategies aimed at facilitating environmental conservation, the management of environmental risks and the creation of new mechanisms for the realisation of values (Jordan *et al.*, 2003). Environmental governance was supposed to be different from traditional environmental regulation, because it focused both on laws and policies and on informal institutions in a search for higher efficiency and more effective organisational structures (OECD, 2011). Moreover, rather than a complete transformation of conventional environmental statehood, the success of environmental governance depended on the re-regulation of conservation and on the use of natural resources, which often combined state-oriented and market-oriented approaches (Mansfield, 2007). In the European Union (EU), for example, environmental regulation achieved a 'mature phase' – according to the governance agenda – after the Single European Act of 1987, which introduced explicit environmental policy provisions and, as a result, the number of Directives and legally binding rules increased significantly (Hildebrand, 2002). The 'mature phase' of environmental statehood was a reaction to the implementation deficit of the previous approaches and recognised the need to reorganise environmental policies in order to make them more dynamic and knowledge-based (Wurzel, 2008).

The experience in the EU represents the best example of the pursuit of governance as the ultimate response to the need to maintain and legitimise the public sector's authority over the interconnections between economy, society and the rest of nature. It could even be argued that the EU took over from the USA in terms of environmental regulation leadership after the golden period of American environmental legislation in the 1970s. EU policies became characterised by a more conscientious association between economic demands and environmental protection with a gradual shift from centralised approaches towards an emphasis on the risks and benefits of more responsive strategies at a local level. The 1972 meeting of the European Council in Paris paved the way for a series of policy adjustments and additional legislation focused on environmental issues. In the 1980s there was growing emphasis on integrated management and pollution prevention. Having said this, it was really the 1992 Treaty that removed trade barriers, including several related to the environment, and promoted further Europeanisation (Liefferink, 1996). Likewise, while the original Treaty of Rome of 1957 made no mention of environmental protection, it was specifically incorporated into the Single European Act of 1987, in the Maastricht Treaty of 1992 and the Amsterdam Treaty signed in 1997. It was the latter that provided the final step in the consolidation of sustainable development into EU environmental policies, as stated in its amended Article 2:

> The [European] Community shall have as its task, by establishing a common market and an economic and monetary union and by implementing common policies or activities referred to in Article 3 and 4, to promote throughout the Community a harmonious, balanced and sustainable development of economic activities, ... sustainable and non-inflationary growth, ..., a high level of protection and improvement of the quality of the environment, the raising of standards of living and quality of life, and economic and social cohesion and solidarity among Member States.

These high level treaties did pave the way for the formulation of a new institutional basis of environmental statehood in the EU. The Third (1982–6) and the
Fourth Environmental Action Plans (1987–92) acknowledged the shortcomings
of previous, top-down approaches, and aimed to harmonise the objectives of the
internal European market and the aims of environmental protection through
integrated policies within what was described as the 'whole production process'.
During the formulation of these Action Plans, the failures of the previous model
of environmental statehood actually meant that the legitimacy and credibility of
EU environmental policies were called into question. There was a clear case for
more effective forms of environmental regulation aimed at combining top-down
and bottom-up instruments (i.e. hybrid or flexible regulatory instruments). A
new and more user-friendly regulatory body, the European Environment Agency,
was established by the European Union in 1990 (and came into force in late
1993) to formally help the member countries make informed decisions about
improving the environment, integrating environmental considerations into economic policies and moving towards sustainability (European Commission,
1989). The Fifth Environmental Action Programme (1993–2000) was a drive
towards more 'effective' and flexible environmental legislation, as well as the
international competitiveness of the EU in globalised markets, whereas the Sixth
Environmental Action Programme (2002–12) maintained the emphasis on
market-based environmental statehood and deepened the calls for higher sectoral
integration. In parallel, the participation of broader civil society was later
enhanced with more complex and holistic pieces of legislation, such as the
Aarhus Convention on Access to Information, Public Participation in Decision-
making and Access to Justice in Environmental Matters of 1998.

The close association between the agendas of sustainable development and
environmental governance and the transition to more accommodating state formations and liberalised economic activities – all directly connected with the new
model of environmental statehood – is schematically represented in Table 4.1.
The conventional and flexible models of environmental statehood are contrasted
here in terms of three main criteria: the formulation of public policies, the basis
of economic strategies, and approaches to the implementation and legitimation
of environmental regulation. The state changed its emphasis from merely enforc-
ing rules and penalties to promoting entrepreneurialism and modernisation,
which were supposed to directly contribute to environmental protection. The
tendency towards more flexible forms of environmental statehood has also
helped to connect the local and the national spheres of environmental regulation.
In the next chapter it will be shown that, in keeping with the predominant tendency towards flexibilisation and interscalar connections, the core anti-commons
tendencies of state action were also expanded from the simple commodification
of nature into the commodification of environmental degradation and conservation (i.e. the formulation of market opportunities such as carbon trading, freshwater markets and forest certification).

The liberalising reforms of environmental statehood were consistent with the
transition from Keynesian economic policies to a new phase where market

Table 4.1 Transition from conventional to flexible environmental statehood

Area of Change	Main attributes	Details of environmental statehood and its associated state-fix	
		Conventional environmental statehood	Flexible environmental statehood
Public policies	Reasoning	Environmental planning to secure basic conditions of production	Environmental restoration and business opportunities
	Policy priorities	Contain environmental degradation, control activities and penalise offenders; increase public services	Profit from environmental protection and restoration; stimulate proactive behaviour; privatise public services
	Preferred scale of action	National (centralised control)	Global and local (decentralisation via centralised agencies)
	Key disciplines	Environmental sciences and engineering	Economics and business management
Economic strategies	The main purpose of economic policies	Economic growth and reduction of inequalities	Economic growth and entrepreneurship
	Economic instruments	Nature commodification via public infrastructure and resource extraction; nominal public service tariffs	Nature commodification via market proxy strategies; cost-recovery tariffs of public services
	Main interventions	Large public infrastructure (e.g. dams and irrigation projects); fixed environmental charges and penalties	Optimisation of infrastructure and management approaches; risk-based environmental charges, incentives and penalties
	Economic outcomes	Exacerbate conflicts; partial coverage of public services	Attempt to contain old and new conflicts; persistent inequalities
Implementation and legitimation	Discursive construction	National development, natural reserves and infrastructure deficiencies	Sustainable development, resource scarcity and infrastructure modernisation
	Implementation strategy	Primarily coordination between state agencies	Multiple coordination between public and private agents
	Social repercussions	Living with risk; sectoral intervention	Coping with risk; multisectoral interventions
	Paradigmatic examples	USA National Environmental Policy Act (NEPA) 1970	EU Water Framework Directive (WFD) 2000

transactions have become the hegemonic metaphor for socioeconomic and socionatural interactions. Instead of a focus on the negative externalities identified by Arthur Pigou (as a main influence behind the previous conventional model of environmental statehood), new economic inspiration was found in the ideas of Ronald Coase and the theorem of free markets as the best mechanism for determining the optimal costs of environmental mitigation. In such a context, market rules became not only fully compatible with environmental protection, but were actually considered to be the most suitable responses to the problems associated with expanding mass consumption and mass waste. The new framework of environmental statehood was also justified through Hardin's anti-commons theorisation about the inadequacy of collective property regimes and the necessary expansion of private property over shared socionatural systems.

Around the world, the search for a more flexible environmental statehood had a contingent relationship with the broader ideological and operational adjustments of the state under the influence of economic neoliberalism. The neoliberalised economy of the last quarter of the century comprised new mechanisms of capital accumulation, including the active dispossession and direct misappropriation of public or collective assets (Harvey, 2005). The far-reaching penetration of neoliberal capitalism, even at the micro scale of neighbourhoods, and its way of encroaching upon non-economic activities, means that fewer and fewer locations and sectors across the planet have been immune to its effects. In the same way, neoliberalised environmental management aimed to connect processes at different levels, from the reform of local institutions to the insertion of the national state in globalised markets (Brand and Görg, 2008). Moreover, the evolution of the state configuration from Keynesianism to neoliberalism was neither linear nor coherent, but always imperfect, variegated and contested (Brenner *et al.*, 2010). It was launched, enacted and constantly contested through dynamic processes of scaling and rescaling that have eventually resulted in the hybrid configuration of state policies as a mixture of liberalised and statist economic initiatives and multifaceted interactions between state and non-state actors. Actually existing neoliberal policies tried to bring about the destruction and reconstruction of previous socioeconomic arrangements, but in practice the state was never able to delegate a number of activities that the neoliberal orthodoxy would have preferred done by the market.

Because the implementation of liberalising reforms was less radical than its idealised formulation, it is also possible to detect a distinct line of continuity between the welfare or welfare-developmentalist strategies and the more recent neoliberalised approaches. For instance, some left-wing governments of Latin America (in countries such as Bolivia, Ecuador and Nicaragua) have articulated an anti-neoliberal discourse that, in practice, allows the simultaneous adoption of distinctive neoliberalising strategies (as in the case of carbon markets, concession of oil reserves and the commercialisation of ecosystem features). It is in this blurred transition from conventional to neoliberalised trends that the symbolism of environmental governance, as well as sustainable development and ecological modernisation, became highly instrumental in consolidating

flexible environmental statehood. Through the rhetoric of environmental governance, coordinated responses between different sociopolitical sectors gradually
turned the responsibility back onto civil society, nonetheless in a way that typically maintained long-established forms of political control under the illusion of
free choice and self-determination. Although it is announced as something uni-
versally advantageous, environmental governance primarily creates novel prospects for the circulation and accumulation of capital, and this has generated a
profound asymmetry of gains and increasing environmental conflicts. Governance has been advertised as something neutral and universal, whereas in practice
it has utilised business models that involve a universe of atomised stakeholders
in the production or preservation of profound asymmetries of gains and beneficiaries. It becomes the ultimate guarantor of specifically codified institutional and
social practices that normalise the extraction and use of natural resources and the
reproduction of socionatural systems and processes.

On the whole, far from being linear and automatic, the transition from conventional to flexible environmental statehood depended on specific national and
sub-national circumstances. In most countries, such as the members of the European Union, the end product was a hybrid environmental statehood framework
that combines dominant elements of flexibilisation with the lasting legacy of
centralisation and top-down regulation. To a large extent, this transition is still
unfolding in the early decades of the twenty-first century, at the same time as
neoliberalised state formations are also being questioned and adjusted. The
complex association between neoliberal strategies and environmental management issues is reflected in this unfinished reconfiguration of the state apparatus
regarding the use of natural resources and the protection of ecosystems. The neoliberalised state was expected to creatively incorporate new regulatory
approaches to cope with mounting socioecological pressures. This need to coordinate neoliberal policies with sustainable development and environmental governance certainly provoked structural and qualitative changes in the state, as will
be discussed in the next chapter. First, however, it is necessary to examine the
conceptual elements of this transition from conventional to flexible environmental statehood, which clearly display a strong influence of the political ideas
of Kant and his attempt to 'enlighten' the political conscience of his time.

The Kantian transition to flexible environmental statehood

The transformation of environmental statehood, since the 1990s, has primarily
responded to calls for the flexibilisation of environmental regulation, more space
for public participation and private sector autonomy regarding compliance with
environmental rules. Flexible environmental statehood emerged out of the dissatisfaction with the weight and ineffectiveness of the conventional model of
environmental statehood. The American experience is a case in point. Only three
weeks after Reagan's election to the presidency, his team started working on the
'deregulation' of what were then seen as environmental statutes "devoid of
policy standards and criteria" and a product of the disproportionate control

exerted by activists over the prior Carter administration (Andrews, 1984: 162). Following the liberal temper of the times, a moratorium on new regulation was adopted in the United States and several regulatory relief initiatives, as well as cuts to the budgets of environmental agencies, were introduced.[1] As much as conventional environmental statehood, in the United States and in the rest of the world, was informed by Hobbesian political thinking, the transition to a more flexible framework – supposedly able to reconcile economic intensification and collective environmental problems – was also associated with debates that had taken place during the European Enlightenment, two centuries earlier, which prepared the ground for the consolidation of the liberal capitalist state. In the case of the transition to a more flexible environmental statehood, the main reference was the political thinking of Immanuel Kant.

Kant's mature life coincided with the phase of national revolutions at the end of the eighteenth century, when social and political instability were potentialised, and then eventually contained, by the then emerging, and later hegemonic, bourgeois liberalism. The Prussian philosopher became not only one of the main interpreters of the bourgeois revolutions, but was also a firm advocate of individual rights and liberties. In spite of that, when considering his impressive philosophical contribution, Kant is not always recognised as a major political thinker. In fact, his name is more often associated with his other celebrated works on epistemology, ontology and morality. One of the reasons is that Kant never actually produced a complete volume on politics equivalent to his most celebrated philosophical books. His political ideas are scattered in many different publications, which nonetheless form a remarkable body of knowledge. With strong positions voiced against abrupt changes in state and politics, Kant was famously considered by Marx to be the German philosopher of the French Revolution, whose political writings on property, rights and morality corresponded to the many aspirations of the emerging bourgeoisie. While a supporter of the civic movement around the 1789 Revolution, Kant firmly condemned revolutionary violence and recommended a more gradual process of change. It is certainly not possible to appreciate the importance of Kant's political thinking without also closely considering his interpretation of freedom and ethics.

An obvious implication of the political ideas of Kant, which is not unrelated to the transition to flexible environmental statehood, is the disapproval of radical, more popular forms of political rupture. The shift towards flexible environmental statehood at the end of the twentieth century reflected a search for economic liberties and a facilitating role for the state in a way that was consistent with Kant's reflection upon the lack of political freedom in the European society of his time. The transition to a new configuration of environmental statehood mirrored the political franchise to the general public advocated by Kant, which was clearly contained by the social institutions of private property and the established political order. In that sense, Kantian liberalism was instrumental, for those in a position of power, in the pursuit of flexible environmental statehood and the related agendas of sustainable development and environmental governance. Flexible environmental statehood was supposed to promote the rational regulation of collective socioecological problems, but always following the high moral ground of the state and according to the strong individualism of the contemporary world. In his text the *Metaphysics of Morals*, Kant (1996) argued that the only innate human right is freedom, defined as the independence from being constrained by the choice of somebody else. Instead of focusing on well-being or welfare, the Kantian state is a true champion of freedom, as much as its ability to pass and enforce legislation has the ultimate goal of guaranteeing individual freedoms. The freedom of all human beings becomes the main principle underlying the state and it comes before the equality of all subjects and their independence as citizens.

> Every action which by itself or by its maxim enables the freedom of each individual's will to co-exist with the freedom of everyone else in accordance with a universal law is *right* ... the universal law of right is as follows: let your external actions be such that the free application of your will can co-exist with the freedom of everyone in accordance with a universal law.
>
> (Kant, 1991: 133)

In a period of revolutionary upheaval and constitutional experimentation (not only in France or Prussia, but in the USA and in many other parts of the Western world), Kant centred his political argument on the notion of freedom and on the expression of reason. In the essay *On Perpetual Peace*, of 1795, Kant gives his definition of a perfect political regime that should be based on independent, private proprietor citizens:

> A *republican constitution* is founded upon three principles: firstly, the principle of *freedom* for all members of a society (as men); secondly, the principle of *dependence* of everyone upon a single common legislation (as subjects); and thirdly, the principle of legal *equality* for everyone (as citizens).... [M]y external and rightful *freedom* should be defined as a warrant to obey no external laws except those to which I have been able to give my own consent.
>
> (quoted in O'Hagan, 1987: 143)

One of the most relevant aspects of Kant's political theories for the purpose of our analysis is that, during his long scholarly career, the philosopher occupied himself with not only the quest for the highest level of correctness, but also with the details of a desired transition to better forms of government. The main concept behind this gradual approximation to the ideal state is his notion of 'provisional rights' in the absence of conclusive rights, that is, in a situation in which the legitimate constitution of definitive rights is still non-existent it is possible to consider preliminary, provisional rights as a mechanism for progressive politics (Ellis, 2005). This movement towards the ideal state would depend both on the recognition of provisional rights and also on the efforts made by all people in the pursuit of the highest good. Kant rejects Hobbes's primary concern with pragmatic, necessarily authoritarian, governance, and distances himself from the romanticism of the following generation, in particular the argument of Fitche, in favour of a gradual, rational reformism. In the vein of other authors in the same century, most notably Rousseau, Kant describes social relations between citizens and the state in terms of a social contract, a tacit, non-written agreement that serves as justification for state power. Instead of self-government and wide-spread democracy, Kant (1991) believes that rational action should be enough for a republican regime that has executive and legislative powers clearly separated.

This social contract and the idea of freedom are achieved through the private property of land and other possessions, which should be respected as the means to express freedom. Kant, going against the economic theories of Locke, does not explain how private property originates, but equally has no problems with the fact that the benefits of property are unevenly distributed across society. Major economic inequalities are therefore no obstacle to the achievement of freedom (which is, nonetheless, typically defined by Kant only in general terms). For Kant, without "property the external expression of liberty has no meaning" (Williams, 2003: 99). Private property is the guarantee of freedom, just as the right to have private property is grounded in the right to freedom. Possession should be established by the collective, general will as a universal system of property rights based on reason. The right to property for Kant consists of a collective agreement, a rational consent among individuals who are logically different and who require the state to consolidate private property (Guyer, 2006). Economic independence is for Kant a key requirement for active participation in politics and for the exercise of freedom. Likewise, it is because of the existence of civil society and the legitimate state – that is, a situation distinct from the 'state of nature' – that people are entitled to own property without the need to be physically in contact with the property, something that Kant (1996: 40) defines as 'noumenal possession' (i.e. the possession in-itself or the possession per se).

The Kantian argument represents a serious challenge to the defence of the sovereign state along Hobbesian lines. Kant's

> concept of liberty embodies what is valuable in Hobbes's understanding of the modern state yet at the same time preserves scope for the kind of liberty that is now defended in our key human rights documents.... Kant's notion of liberty involves others but he is not collectivist. There are three main institutions he connects intimately with freedom: property, civil society and a republican constitution.
>
> (Williams, 2003: 99)

In the essay *Against Hobbes*, Kant (1991: 73) even makes an important distinction between the existence of several social contracts and the overarching civic constitution that must be observed by all. While for Hobbes, humans enter a political union motivated by self-preservation, for Kant political union is the dictate of practical reason and the acceptance of authority is the expression of a sense of morality. According to Hobbes, the natural equality of people is not translated into an equal social and economic status for everyone; for Kant, on the contrary, individuals are equal from both the perspective of natural rights and their equal status under the rule of the sovereign. They are not only equal, but able to help to solve the problems of state institutions. Even 'bad men' can do it, due to their intellectual ability, the use of reason and the fact that nature binds people to the will of the state (Saner, 1973). Nature is a source of human fulfilment, but the powers of nature also play a part in the actions of individuals and states. Kant hopes for a unity between "nature and freedom according to the inner principles of the law in mankind" (Saner, 1973: 48). Instead of the focus on the 'state of nature' of Hobbes and his advocacy of a repressive state in charge of the commonwealth, Kant associates human action with the balance between practical and pure reason, that is, between the knowledge of phenomena and the knowledge of things that are good in themselves.

Kant (1991: 84) explicitly criticises Hobbes for placing the head of state above any contractual obligations towards the people and allowing them to act towards the people "as he pleases". Political rights for Kant are based firmly on reason and the state should operate on behalf of society for the achievement of higher levels of reason. The state is thus restricted to securing broadly defined rights, through the legitimate use of coercion, and maintaining only a formalist understanding of freedom, which reveals the Enlightenment's superficial conceptualisation of freedom among equals in the face of huge social inequalities. The action of the state should reflect the need for juridical, harmonious conditions in which rights are the outcome of reason. Instead of seeing the state as an agent of private property, as previously claimed by Locke, there is a mutual dependence between the Kantian state and private property (Kersting, 1992). According to Kant, the state should not be paternalistic in terms of dealing too closely with human needs and interests, but its main responsibility is the preservation of freedom and self-sufficiency (achieved through universal private property). Individual freedom, including property rights, must be restrained by the state in a way that leaves the same freedom for everyone. A clear example of the clever, although restricted, flexibilisation that permeates his political thought is the fact that Kant (1996) contemplates the role of a 'permissive law of practical reason' to complement the conventional law based on command and prohibition.[2]

Kant's political thinking does indeed help to understand the achievements and the deficiencies of the institutional changes associated with the evolution from Hobbesian-inspired environmental statehood (focused on the knowledge and authority of the state) into more responsive and dynamic ways of dealing with collective issues, that attempt to bring freedom to the centre of politics and to combine pure and critical reason. The Kantian elaboration on political freedom, provisional rights and universal laws had direct ramifications for the transition to less rigid and more widely supported environmental politics. From a political ecology perspective, two main elements of the Kantian argument are particularly relevant to the transition towards flexible environmental statehood, namely, the idealised liberal state and the attempt to reconcile several dualisms (especially the division between phenomena and the thing-in-itself). These two key aspects of Kant's political theory were particularly relevant, even indirectly, during the transition from a rigid model to a more flexible framework of environmental statehood.

First, according to Kant, and contrasting with Hobbes, no one should be entrusted with absolute power because even the best-informed and well-meaning of people will be subverted by power. In his opinion, *"[w]hatever a people cannot impose upon itself cannot be imposed upon it by the legislator either"* (Kant, 1991: 85). Nonetheless, this does not mean that people should be entitled to select their government and representatives through vote. On the contrary, for Kant, the basis of state power is neither public representation, nor the welfare of citizens, given that the state cannot impose particular conceptions of happiness upon its people, and therefore freedom should be the only universal aspiration. Kant sees the state not as an impediment to freedom, but as the means for freedom, as long as it acts rationally and respects the limits of freedom. Instead of paternalistic calls for happiness, Kant devises a strong state – centred on the innate right to freedom and promoter of superior reason – that should be capable of maintaining rights and avoiding authoritarianism. The Kantian elaboration represents an attempt to steer a middle course between theoretical and practical elements of human agency, between the idealisation of freedom and reason and the obstacles to their achievement in the real world. His philosophy is located in the middle ground between the rationalism of Leibniz and his followers, and the empiricism of David Hume and other British thinkers. Kant wants all individuals as co-participants in the government of collective matters, but relieves the state from democratic elections and effective popular control.

The liberalism of Kant is obviously an advance in relation to the absolutist regimes of his own time, but its key limitation is to advocate freedom from the perspective of those already in command of the state and, above all, for the protection of the institutions of private property (which was needed for the capitalist modernisation and industrialisation of the world). The social contract that unites individuals in general would make the state the legitimate guarantor of freedom to all members of civil society, but there is nothing to guarantee that the state is freely chosen and genuinely legitimised by the individuals it is meant to liberate. Kant doesn't eliminate the possibility of welfare or well-being legislation (Kaufman, 1999), but limits the scope of such laws and policies to the canon of conservative liberalism and capitalist modernity. Crucially, it is precisely because of the narrow configuration of democracy and freedom (i.e. within private property boundaries and commanded by rationalism and rational choices) that Kant's political philosophy provided, two centuries later, a fitting justification for the adjustment of environmental statehood according to the intricate demands of a globalised market-based society. Kant's conception of the liberal state in relation to private property rights was particularly relevant for dealing with environmental resources threatened by overexploitation and difficult to control through rigid command-and-control regulation, such as marine fisheries and rainforest destruction (Breitenbach, 2005). This defence of a liberal social contract made Kant the most suitable philosopher for the transition to flexible environmental statehood (albeit with a lesser impact on its consolidation, which will be discussed in Chapter 6).

If the Kantian proposal sets limits to the coercive functions of the state – with the state able to enforce environmental regulation and thereby constrain the action of citizens and economic operators to a certain degree, but without seriously affecting the freedom of private property – his deontology tries to provide the moral imperatives for personal behaviour that is environmentally more responsible. Kant aims to associate practical reason with moral consciousness or moral reasoning in the form of the 'categorical imperative' of universal laws. However, the problem remains of how to follow such universal laws in a world fraught with sociopolitical inequalities and a highly asymmetric balance of power. Especially in the recent history of the European Union, it is possible to find clear repercussions of Kant's ideas on political transition and rational political institutions on the transition from the rigid and onerous environmental regulation into more sophisticated, light-handed approaches. By placing the state as the ultimate driver of individual freedom and social morality, Kant validates more responsive, supposedly amicable public policies based on rational assessments and interaction between (formally) equal citizens. Kantian-informed policies prevailed in the 1990s, especially after the Rio Summit and the introduction of new legislation and guidelines. Interestingly, the aforementioned Fifth Environmental Action Programme, in place since 1993, received the suggestive title 'Towards Sustainable Development' and made it clear that a different direction of environmental regulation and policy-making was needed. It was recognised in the document that

> while a great deal has been achieved under these programmes and measures, a combination of factors calls for more far-reaching policy and more effective strategy at this juncture.... The achievement of the desired balance between human activity and development and protection of the environment requires a sharing of responsibilities which is both equitable and clearly defined by reference to consumption of and behaviour towards the environment and natural resources.
>
> (European Commission, 1993: 11)

Second, Kant's epistemology complements his ideas about the transition towards an idealised state. For Kant, aprioristic reason is essential for the understanding of the connections between all the elements of an object or of a phenomenon. Through the comprehension of their relationships with the world, humans can become 'legislators' of nature, that is, mediators of actual phenomena; and it is in such a process that 'reason reasons', in the sense that with a priori concepts it is theoretically possible to attain a middle term in relation to understanding (Deleuze, 2008). Kant famously reworked the philosophical debate on the possibilities of analytic and synthetic reasoning, that is, combining the dogmatist claim that through theoretical reason it is possible to go beyond experience and the empiricist argument that it is not possible to go beyond experience. Kant concludes that synthetic judgments can be made a priori, as in the case of mathematics (e.g. $7+5=12$, but 12 is neither in 7 nor in 5). That leads Kant to advocate new bases for metaphysics and to prioritise practical reason as the means to know things as they are. It is significant that Kant also takes into account the position or the circumstances of the knower, as an influence on the understanding of the world. For Kant, whereas the real world, the world of things-in-themselves (the noumenal realm), cannot be apprehended, the world of appearances cannot be explained by experience alone (the realm of phenonema). Consequently, the understanding of the world requires logical reasoning and prior abstraction, or in his words, the experience "teaches us that a thing is so and so, but not that it cannot be otherwise" (Kant, 1929: 43).

In that sense, the interpretation of the lived world would require a priori synthetic concepts that result from human intelligence. Kant (1929: 45) combines experience with the formulation of pure, aprioristic principles that are "indispensible for the possibility of experience". Even common understanding is permeated by, and therefore 'never without', certain modes of a priori knowledge (Kant, 2002). In addition to the unresolved reconciliation between pure and empirical reason, the Kantian epistemological system incorporates several other conceptual dualisms such the tension between freedom and coercion, general rules and individual rights, public good and private property. At the centre of those dualisms – typical of the Enlightenment period and with major repercussions for Western society ever since – there is the perennial contrast between the ideal, untouchable world and the phenomenological world of experience. This Kantian dichotomy between pure and practical forms of thought was also vividly present in the transition to a more responsive and flexible environmental statehood, in particular with its normative calls for environmental conservation detached from the socioecological problems actually being experienced and the associated levels of injustice. Against the alleged rigidity and ineffectiveness of the previous conventional approaches, flexible environmental statehood was set on the course of creative, responsive strategies aprioristically informed by concepts such as governance, ecological modernisation and sustainable development. As a result, the contemporary (capitalist) state is an 'aprioristic' agent that formulates responses to environmental problems (which are put forward as necessary and universal) and then attempts to rationalise its application based on those preconceived postulations.

The defence of aprioristic responses to concrete problems was effectively one of the hallmarks of the transition from conventional to flexible environmental statehood, which also helps to explain the deficiencies and contradictions of this transition. The reform of environmental statehood emerged from the central Western states as a priori, pre-given solutions that had to be reinterpreted and re-rationalised along more flexible lines. Possible alternatives to collective environmental problems should emanate from a priori judgments, that is, of what is considered necessary by the apparatus of the state. Nonetheless, the inherent dualisms of the Kantian argument had the evident side effect of failing to provide appropriate justification, even for the purposes of the conservative political and economic sectors, for global environmental governance and flexible environmental statehood (this gap was later resolved, at least temporarily, by resorting to Hegelian political theory). Following Kant's dichotomy between the two realms of reason – aprioristic reason and practical reason – the deeper barriers to the resolution of socioecological questions were only superficially addressed and largely preserved during the transition to flexible environmental statehood, meaning that the contradictory agenda of pro- and anti-commons that had permeated environmental statehood remained practically unchanged. Furthermore, the individual remained split between two realms, one associated with reason and the other connected with experiences and senses. Kant's answer is for people to act in a way that personal action should become a universal law, but also to treat other human individuals as an end and never as a means. This abstract sense of 'duty' is quite problematic and was later criticised by Hegel because of its 'emptiness' or lack of content and detachment from the concrete social order (O'Hagan, 1987).

An aprioristic treatment of environmental problems, according to the Kantian advocacy of reason and morality, proves to be highly convenient when reaffirming universal freedoms at the edges of bourgeois liberalism and private property liberties. The morality of private property and the centrality of aprioristic approaches ultimately reflect the attempt to expand, through a flexible environmental statehood, the capitalisation of the world and the commodification of the many still untouched socionatural processes and resources. The commodification of socionature is not merely a material procedure, but it requires first of all the aprioristic conversion of the multiple values of socionature into economic, monetised translations of value. In other words, the commodification of additional features of socionature is a logical expression of the Kantian synthetic apriorism, in the sense that the capitalist knowledge of the world rests upon principles that are supposedly self-evident (in this case, the idea that commodification and marketisation of socionature would foster efficiency and environmental protection). With the commodification of socionature, a primacy of the exchange-values is established at the expense of use-values, and a range of commercial transactions can take place, as in the case of paying for ecosystem services and the privatisation of public water utilities. The quantitative character of exchange-value (in contrast with the qualitative basis of use-value) became strategic for the reduction of the plurality of value symbolisms into the unified vocabulary of market relations that underpin the neoliberalisation of nature. Marx (1973: 221) had already observed that

> before it is replaced by exchange value, every form of natural wealth presupposes an essential relation between the individual and the objects, in which the individual in one of his aspects objectifies himself in the thing, so that his possession of the thing appears at the same time as a certain development of his individuality.

The conversion of the multiple values of nature into the narrow grammar of commodification is part of the simplification of socionatural interactions that follows the accumulation needs of contemporary economic relations. The politics of commodification became part of the appropriation of socioecological systems according to the demands of capital accumulation and the broader agendas of political control by the neoliberal state apparatus (discussed in more detail in Chapter 5). The tensions around the commodification of things and processes are neither static, nor linear, but constitute an arena of constant disputes over existing social structures and cultural identities (Appadurai, 1986). Rather than a simple phenomenon, the commodification of nature is more than an economic relation. The forces of nature commodification play an important role in recognising politico-economic subtleties and the collective negotiation of the meaning of nature. The process through which 'goods' or 'materials' may become 'commodities' is not straightforward and unequivocal, but complex, varied, sometimes unpredictable and enigmatic (Appadurai, 2005). A commodity cannot be described as simply a physical and material object, but it is, in fact, the form and the social relations around the commodities that confer its character. As emphasised by Lefebvre (1972: 98), "commodities do not assert themselves qua things but rather qua a kind of logic" that is permanently negotiated.

All things considered, the Kantian doctrines on freedom, reason and his aprioristic epistemology aided the transition to more flexible environmental governance, although they also left many questions unresolved and in need of further adjustments. The combination of ideas about a controlled state liberalisation, aprioristic thinking and moral reasoning represented an apt recipe to improve environmental statehood according, primarily, to the needs of politico-economic hegemonic sectors. Kantian thinking was consistent with state reforms according to the goals of environmental governance and sustainable development as an attempt to mediate the degradation caused by the intensification of socio-economic activities, without imposing radical changes to the patterns of production and accumulation. As a result, the evolution from government to governance could be seen as the expression of political equality and freedom while it also justified, and helped to reinforce, economic inequality. Deleuze (2008) points out that Kant seems to strike the right balance between extreme rationalism and empiricism, which is precisely what makes him appropriate to inform the transition to flexible environmental statehood (i.e. mixture of moral claims of sustainability and more immediate adjustments based on practice). Nonetheless, the Kantian state theory is seriously undermined by his idealisation of the possibilities of state action and the maintenance of structural dualisms allowing the renewed exploitation of socionature. Kant's political formulation replicates the fundamental paradox of the capitalist encroachment upon socionature: how to reconcile anti-commons, private interests with the long-term needs of wider society and of the rest of socionature. A concrete example of the problematic transition from government to governance, in the context of Kantian-inspired reforms, is discussed below (please note that this section is directly related with the case study presented at the end of the previous chapter).

The transition to more flexible environmental statehood in Brazil

The Brazilian experience of flexible environmental statehood is a case in point of the inherent limitations of the institutional transition informed by Kantian calls for reason and freedom. The redesign of the Brazilian public sector started in 1995 with the publication of the 'White Paper on The Reform of the State Apparatus', which included a new set of criteria for investing in infrastructure and the management of public utilities. The justification was, on the one hand, the lack of public funds to modernise and expand public services, and, on the other, the supposedly ineffective and wasteful operation of state-owned enterprises. Early signs of environmental statehood in Brazil began with the approval in 1934 of the Brazilian Forest Code and Waters Code, as well as legislation in the same period that established measures providing economic assistance to natural rubber producers (Ioris, 2007). More specific environmental regulation followed the approval in 1981 of the National Environmental Policy Law, as the first Brazilian statute to include a legal definition of the environment: "the set of physical, chemical and/or biological conditions, laws, influences and interactions that facilitate, shelter and govern life in all of its forms." This legislation created the National System for the Environment and the National Council on the Environment, which were clear indications of the implementation of conventional environmental statehood in the 1970s and 1980s. The shortcomings of that institutional model resulted in several legal and institutional adjustments adopted in the 1990s, which reoriented the direction of environmental statehood increasingly in favour of collaboration with and incentives to private business sectors.

In this context, the reform of the Brazilian water sector offers a very emblematic illustration of a Kant-informed transition to a more flexible environmental statehood, still within the anti-commons aims of the capitalist state. The overall reorganisation of the public sector had direct repercussions for water regulation and water management. Within the structure of the Ministry of the Environment, a new water secretariat was created in 1995 in charge of the coordination of national policies and other legal reforms under debate in the parliament. With the approval of the new legislation in 1997, the National Water Resources Management System (SINGREH) was established to bring together various public agencies and consultative committees. The structure was completed in 2001 when the National Water Authority (ANA) was installed to be responsible for water use permits and the implementation of technical programmes. The legislation introduced new regulatory instruments, such as plans, river classification, licences and bulk water charges, which are classical tenets of the governance canon mentioned above. Although the new regulatory context encouraged the formation of 'multistakeholder approaches' that were supposed to involve all the social actors, the paternalistic forms of public engagement meant that participation could be operationally constrained by formal and informal bureaucratic goals, particularly when oriented towards concerns that were external to the local reality. The approval of plans and the reconciliation of spatial differences were delegated to catchment committees, yet the core element of new policies was, and still is, the expression of the monetary value of water.

It was certainly not a coincidence that the introduction of new water management institutions happened together with the liberalisation of the Brazilian economy, which consisted of declining public investments, high interest rates, labour market reforms, high unemployment and attraction of foreign investors. In that context, newly formed decision-making forums were dominated by the same rural and urban oligarchs that traditionally controlled economic and social opportunities related to water use and conservation. Instead of promoting a genuine change in public policies, the new approaches largely preserved the hegemonic interests of landowners, industrialists, construction companies and real estate investors, at the expense of the majority of the population and ecological recovery. The administrative structure of water regulation, which extended from the federal government to state authorities and river basin committees, achieved only marginal results in terms of environmental restoration and conflict resolution. This appropriation of the Brazilian State by the stronger economic groups corresponded to the policy distortions described by Lefebvre (2009: 58–9): in the countries where the state is the main force of economic growth, the state apparatus itself becomes both the site and the stake of social struggles; there is a real possibility of "the formation of a bourgeoisie that would not be a trading or commercial bourgeoisie, but a bourgeoisie linked directly to the State apparatus, a bureaucratic bourgeoisie, that is to say, entirely new social formation".

A careful examination of the first years of the new legal framework reveals disappointing results in terms of reducing impacts and improving the management of water systems. This gloomy picture is formally acknowledged by the Ministry of the Environment (MMA, 2006), in particular the widespread sources of pollution in urban areas (e.g. only 47 per cent of the municipalities have sewerage systems and only 18 per cent of the total sewage is treated) and in the countryside (e.g. around 70 per cent of the watercourses between Rio Grande do Sul and Bahia are polluted by agro-chemicals used in intensive crop production). In addition, resource availability has been compromised by the over-extraction of water and the continuous construction of large dams. The hydropower projects approved by the national administration, in spite of strong public opposition, clearly evidentiated the priority of 'economic growth at any price'. The overall trends of water degradation and, more importantly, the selective involvement of the public in the decision-making process seem to suggest a more fundamental weakness in the ongoing water reforms. The failures of the new water policies actually suggest that the theory of integrated water management has been mechanically pushed through by multilateral agencies to grant functions to a system yet to be constructed (Abers and Keck, 2006). The legal reforms have prized the influence of private agents in the formulation of water projects (e.g. hydropower schemes and public water companies), which at the same time has raised novel opportunities for capital accumulation via, currently, the adoption of ecological conservation measures.

From local to national initiatives, the acceptance of the imperative of development in Brazil has remained a strong feature of water policy-making under flexible environmental statehood (for instance, new techniques developed for the assessment of water projects maintained that the design of new hydropower schemes should include the environment as merely a 'variable in the equation'). The recent experience has shown that politicians are always too keen to force the authorisation of new public or private initiatives on the grounds of raising tax revenue and job creation, even when the actual result is evident and widespread social and ecological disruption. For instance, in 2005, the Ministry of the Environment was forced to approve a questionable project for water transference from the São Francisco River to northern catchments in the semi-arid region. This inter-basin project has been vehemently criticised on the grounds that the benefits of water transference are likely to be appropriated by political leaders at the risk of socionatural impacts to both the source and the receiving catchments. Likewise, in 2007, the same Ministry was compelled to grant licences for the construction of two large hydropower schemes along the Madeira River, in the heart of the Amazon region, regardless of direct disapproval from its senior staff and technical experts. The hasty approval of the large Belo Monte dam in 2012, to be constructed along the Xingu River, on the Eastern side of the Brazilian Amazon, despite the existence of comprehensive water and environmental legislation, is another clear demonstration of the failure of environmental statehood to deal with the perverse balance between local socioecological impacts and economic benefits (i.e. in this case, low cost energy made available to other regions, São Paulo in particular). In other regions of Brazil, hydraulic projects continued to be approved and implemented even if they violate traditional community rights over common resources (Ribeiro *et al.*, 2005).

It is relevant to observe here that the very first article of the 1997 water law established the primacy of neoclassical economic theory over water management in Brazil. The article recognises that "water is a scarce natural resource, which has economic value". There is here an unambiguous resemblance to the fourth UN Dublin Principle (approved at the 1992 International Conference on Water and the Environment) which declares that "water has an economic value in all its competing uses and should be recognised as an economic good" (see more in Chapter 5). This phrase encapsulates the two fundamental tenets of the neoclassical economic paradigm behind flexible environmental statehood: the idea of a scarce resource and the (economic) value of water. In effect, the expression of the economic value of water has been the main concept supporting the formulation of subsequent policies and initiatives in the last decade in Brazil. As repeatedly mentioned in the official publications, because of its quantitative scarcity and declining quality, water is no longer a 'free good', but has clear economic value. In other words, because water is (or was made) scarce, it now requires an economic treatment to address existing and future problems. Once the monetary value of water is determined, it can be managed as any other economic factor of production that has marketable costs, effects and benefits. The most relevant expression of the monetised value of water in Brazil has been the imposition of bulk water user charges (i.e. 'water pricing') under the 'user-pays principle' or the related 'polluter-pays principle'. According to the mainstream economic approach, those wanting to extract surface and ground water or dilute effluents in the watercourses should pay a charge proportionate to the negative impacts caused (i.e. environmental externalities). For example, in the Paraíba do Sul the charging methodology demands that all water uses above a certain threshold (i.e. consumptive uses above one litre/second and hydropower bigger than one megawatt) must pay a monthly fee, calculated by taking into account three factors: the extraction rate, the percentage of use and the quality of the effluent. There is a standard charge (R\$0.02/m^3) for industries, water supply and mining, and significant discounts are offered for agriculture and aquaculture in this particular river basin (Ioris, 2009).

The introduction of bulk water user charges in Brazil aimed to minimise social costs through the determination of the optimum scale of operation, induce rational economic behaviour and generate revenues for environmental restoration and law enforcement. However, since the early days of the new regulatory regime at the end of the 1990s, the imposition of water charges has caused controversy on national and local scales. In many catchments, the political maelstrom related to the controversial introduction of water charges has hijacked the broader debate on environmental restoration and prevention of impacts. The perverse consequence of water user charges is evident in the areas where it has already been adopted, in particular the split of stakeholders into confrontational groups and the widespread suspicion about hidden sector agendas (as in the Paraíba do Sul). Such new water policies, including the introduction of water pricing, need to be understood in the broader context of national and regional development influenced by neoliberal demands. Market-friendly water policies were particularly important to agroindustrial companies that move their activities to water-scarce areas (examples of these are industrial parks and the irrigation projects being constructed in northeastern states, such as Bahia, Ceará and Pernambuco, and agribusiness complexes in the states of Goiás and Mato Grosso). Because the companies were attracted to the areas of economic frontier due to the availability of natural resources, cheaper labour force and fiscal incentives, the payment for water user charges was a matter of only secondary importance. Rather than a simple fee, paying for bulk water use represented an additional guarantee that the conditions of production (water availability in this case) would be provided and maintained by the state. However, the combination of water and development policies ended up causing a twofold penalty to traditional water users, given that these are usually less well prepared to cope with the new water charges and, more importantly, their activities may be seen as second-rate when compared with those of the newcomer water users.

Instead of improving the environmental condition of river catchments, the payment of bulk water charges tends to be tacitly used to validate the operation of environmental impacting activities: the payment for water use by industries, electricity operators and irrigators is utilised as a political justification for avoiding close scrutiny. That has been the case with industrial effluent discharges in the Paraíba do Sul catchment, where the industrial sector has been able to preemptively manipulate the approval of water charges to suit their demands for soft regulation. At the same time, larger industries have opportunistically used their payment for water use to improve their commercial image as corporately responsible (Féres *et al.*, 2005). Since industries were officially involved in the new water regulation, there was scarce room for calling into question their responsibility for the poor environmental quality of the catchment. In spite of the 'inclusive negotiation' that, according to Formiga-Johnsson *et al.* (2007), characterises the local experience, there has been a formal acknowledgement by the local stakeholders that the implementation of water charges has not progressed as expected. On the contrary, the introduction of bulk water charges has contributed little in terms of environmental restoration in the Paraíba do Sul: the official statistics show that, between 2003 and 2006, the charging scheme was responsible for collecting a total of R\$25.4 million, which is considerably less than the estimated amount needed to restore the catchment (i.e. an annual investment of R\$360 million or R\$4,600 million by 2025, cf. Coppetec, 2006).

Another significant element of the conservative 'modernisation' of the Brazilian public sector was the programme of public utility privatisation. The privatisation of electricity and basic sanitation companies represented around a quarter of total assets transferred into private hands (approximately US\$100 billion were transferred into private hands, either through full divestiture or through operational concessions of public utilities).[3] Because 90 per cent of the electricity generated in Brazil comes from hydropower schemes, the privatisation of energy has in effect been an indirect form of water resources privatisation. So far, most of the electricity distribution companies and around 40 per cent of the generation companies owned by the state have been sold off to private operators. In nominal terms, the transfer of electricity companies to private hands attracted US\$23.5 billion (Anuatti-Neto *et al.*, 2003). Around 48 per cent of the payments made by private investors to acquire electricity companies were remarkably financed by government-owned banks (particularly via the national development bank BNDES). The involvement of private operators was also facilitated by changes in the legislation that removed the difference between domestic and foreign firms. Privatisation was further encouraged by reducing investments in public utilities prior to the sell-off (i.e. to reduce political opposition due to the deteriorating performance of state-owned utilities), contractual clauses that protected privatised companies against changes in the exchange rate, electricity tariffs rising above inflation and the removal of compensatory subsidies to low-income families. Since 2003, the federal administration has reduced the emphasis on the full divestiture of public electric utilities, but has maintained other traditional options of private sector involvement by contracting out services and public–private partnerships.[4]

In contrast to the hydroelectric sector, the privatisation of water supply and sanitation has been more restricted and has faced higher political resistance. Because of lengthy negotiations and legal disputes, only 3 per cent of the water supply and sanitation utilities were privatised, which serve now around 5 per cent of the national population (Britto and Silva, 2006). One fundamental obstacle is the hybrid responsibility that characterises water services in Brazil: according to the federal constitution, municipal authorities are in charge of water services, whereas the great majority have delegated the operation to companies owned by the state (provincial) governments. The agreements between municipal and state authorities were formalised in the 1970s, during the military dictatorship, when the national policy was to concentrate resources and power in the state utilities. Under the influence of the liberal policies of the 1990s, some state administrations dissolved or demobilised their water companies, unilaterally returning the responsibility to the municipal administrators. This allowed some municipal administrators to transfer the local water services to private companies (including many foreign ones). Privatisation was further encouraged by the reduction in investment by the central government, which is responsible for managing the main investment fund (i.e. FGTS)[5]: between 1995 and 1998 only around US$1.0 billion was invested in the sector, while US$4.0 billion of past loans were paid back to the central government. This meant that a surplus of around US$3.2 billion was retained in the investment fund, regardless of the urgency of social demands (Oliveira Filho, 2006). During this period, a specific agreement was signed with the IMF committing the Brazilian government in 1999 to broaden the scope of the privatisation of water services. The result is that the average annual public investment between 1995 and 1998 totalled US$380 million, but the same average reached only US$38 million between 1999 and 2002 (it was zero in 2001). In parallel, while the central government reduced the access of public utilities to governmental funds, incentives and loans were made available to attract the attention of private operators.

The privatisation of water supply and sanitation under neoliberal pressures was only one element of a very complicated sector that failed to serve 24.2 per cent of the population with drinking water and 46.2 per cent with sewerage services (IBGE, 2004). Profound injustices have been inflicted upon marginalised social groups who, especially in the larger cities, have been forced to live in floodplain areas prone to flooding and lacking the most basic water infrastructure. In many cases, such as in the metropolitan area of Rio de Janeiro, poor households only have access to precarious water services and need to complement that with the purchase of costly water from private vendors. If in the previous decades water supply and sanitation were restricted to the wealthier cities and neighbourhoods, this recent privatisation of publicly owned companies has done little to improve the situation. Instead of higher investments and efficiency, privatised companies have been criticised for charging more for a worse, less reliable service. In many situations, privatisation has shifted "the burden for providing services to the poor from society as a whole and back to the poor themselves" (Mulreany et al., 2006). Privatisation has also raised a range of conflicts between private operators, public regulators and customers, as well as evidences of corruption and wrongdoings. The concession process has been far from transparent, despite steady increases in tariffs and charges (for instance, the charge to connect to the water network system in the city of Limeira increased from 65 per cent to 176 per cent of the official minimal monthly salary after privatisation, with no discounts for low-income families, cf. Vargas, 2005).

Similarly to what happened in other countries, utility privatisation in Brazil has faced significant scepticism about the real motivations of private companies that are more accountable to the shareholders than to their customers and, at this point in time, the future of the water sector is uncertain, with unclear legislation and ambivalent policies. Nonetheless, the reaction against utility privatisation has been mixed and sometimes hesitant. On the one hand, grassroots organisations have worked together with the National Association of Municipal Water and Sanitation Utilities (ASSEMAE) to demonstrate the importance of maintaining both the ownership and the operation of water companies in the hands of the state. ASSEMAE represents more than 2,000 Brazilian municipalities that have a direct administration of their water services. The remaining municipalities (approximately another 3,700 towns and cities) have delegated the water service to state water companies or private operators. This movement against privatisation has underscored good examples of publicly managed services, such as that of Porto Alegre, where the combination of autonomous public mobilisation and a competent left-wing administration (Heller, 2001) achieved an enduring transformation of the services provided by DMAE, the municipal water utility (Holland, 2005). On the other hand, while new legislation on basic sanitation was passed in the year 2007 (Law 11,445) which emphasises the provision of water services as a basic human right, it also encourages the formation of 'public–private partnerships' as an important strategy for improving and expanding water services. These partnerships have been tacitly used in Brazil and in other countries as a disguised form of utility privatisation (see Chapter 5). There are also signs that increasing the participation of the private sector in the provision of water services (not necessarily through privatisation, but also via other flexible business arrangements) has attracted growing sympathy even from ASSEMAE members and left-wing politicians. The alternative to 'public–private partnerships' often mentioned by activists is the formation of 'public–utility partnerships' (also called 'public–public partnerships'), but sometimes these are promoted in the wake of failed private-sector contracts, and the result is the cherry-picking of the most profitable areas and the neglect of less profitable communities (Hall and Lobina, 2007).

Apart from the 'modernisation' of the public sector and monetary valuation, the market-based solutions that have underpinned the institutional reforms in Brazil have increasingly facilitated the adoption of other indirect mechanisms of water commodification. One of these new forms of converting nature into tradable commodities has been the payment for ecosystem services (PES), which includes 'services' related to watershed conservation such as the maintenance of clean water supply and protection against soil erosion (Kosoy *et al.*, 2007). The rationality of PES is directly inspired in the neoclassical concept that free market operations can guarantee the most efficient solution to environmental externalities. The justification is that those who benefit from the ecosystem services should be prepared to make direct payments to the local people more closely associated to the conservation of the ecosystem. For instance, if the protection of an upstream forested area helps to maintain river flows, the environmental service (in this case, the guarantee of water availability by protecting the forest) should be paid by downstream water users. PES entails a full interchangeability between the market inputs used by the industries and agriculture and the non-market service of maintaining the river flow. The first requirement before PES can be adopted is obviously the estimate of the monetary value of the environmental services. The calculation is normally processed through ecosystem valuation methods, which normally produce significant inconsistencies. For example, Fearnside (1997) estimated that 10 per cent of Brazilian agriculture depends on rainfall originating from the evapotranspiration in the Amazon, which would correspond to an environmental service (i.e. guarantee of rainfall) that is worth US\$7 billion per year for the entire rainforest.

Many Brazilian academics and policy-makers have embraced PES as a very ingenious option for dealing with water management problems. The National Water Authority (ANA) launched the ambitious 'Water Producer' programme, an initiative that offers financial compensation for soil conservation interventions that potentially increase or maintain water availability. One of the catchments covered by the programme, located in the municipality of Extrema, contains a significant proportion of the freshwater supply to the city of São Paulo and, in 2007, landowners started to receive financial support to adopt soil conservation measures that indirectly protect watercourses. Another similar initiative was the Catchment Pollution Removal Programme (PRODES), which 'buys' the treatment of sewage by private or public operators (instead of the direct financing of the sewage works). The attractiveness of PES was also demonstrated by two 'private members' bills' recently introduced and under discussion in the National Congress (bill 142/2007 in the Senate and 792/2007 in the House of Representatives). Similar propositions were presented in various state assemblies to further regulate the payment for ecosystem services in areas under local jurisdiction (e.g. in the State of Acre). For many academics and politicians, the win-win promise of PES seems the ultimate proof of the perfection of the market, which is capable of finding inventive solutions to the very problems it causes. In their view, PES not only introduces a 'sophisticated' response to environmental degradation, but also generates new commercial opportunities related, for example, to the certification and monitoring of environmental services.

On paper, the certification of environmental services seems to have the ability to promote environmental protection, since water users would become more aware of the economic value of ecosystems. In practice, however, the success of PES in terms of protecting and restoring the environment has been close to nothing. The disappointing outcomes of the PES experience can be explained by various operational and conceptual frauds. First of all, it is extremely difficult to relate the provider of the service with those willing to pay for it. The adoption of PES has also been hindered by demand-side limitations and a lack of supply-side know-how (Wunder, 2007). Second, PES only works in situations where the threat of environmental degradation is extremely high. This is because it requires an irrefutable proof of the environmental risk to persuade beneficiaries to accept the payment for the service. If the PES regime becomes more widely adopted, it can even induce the artificial 'fabrication' of environmental threats in order to justify the payment. In other words, the implementation of PES can divert the attention away from environmental protection towards profitable market transactions. Third, in the few cases where it has been adopted, the price of the environmental service is not the outcome of free market bargain, but on the contrary, is created by the regulatory demands and opportunistic behaviour of private firms (see Robertson, 2007). Fourth and more importantly, the market logic behind PES is fundamentally based on a utilitarian relation between people and nature that ignores the capacity of local populations to appreciate the value and spontaneously protect their ecological base. This rationale assumes that human beings are naturally inclined to convert natural resources into cash and, therefore, people need to be paid to avoid causing environmental harm (this is, for example, the argument of Vosti *et al.*, 2003, for the protection of the Amazon Basin). It overlooks the fact that local populations have a long history of skilful interaction with the environment and that the pressures over natural resources are, to a large extent, created by economic globalisation, the same globalisation that now encourages the adoption of artificial schemes like PES.

Interestingly, there are many similarities between the intense influence of mainstream economics on the reform of the water sector in Brazil and the comparable experiences in the majority of Latin American countries. New forms of dealing with water management in the region started to be implemented in the region after the end of the military dictatorships in the 1980s, when the approval of liberalising laws to regulate environmental conservation and utility operation coincided with a whole range of market-friendly measures. This included the closure of government departments, the privatisation of government-owned assets and the aggressive appeal to foreign investors. The commonalities between the Brazilian experience and what happens elsewhere in Latin America is not simply a coincidence, but attests to the exogenous origins of the recent water reforms. In the same way that development banks and multilateral organisations encouraged the expansion of water infrastructure after the Second World War, the current water reforms are fundamentally grounded in concepts that emanate from universities and think tanks based in the North. This is a main reason for the systematic difficulty to connect local demands and the values of local populations with the language and the targets of the centralised regulatory agencies. Even with a large proportion of the freshwater available on the planet, the water sector in Latin America merely reproduced the trends in investment and reorganisation imposed by the leading economies. Such reforms have not happened in a vacuum, but were in fact intimately related to the patterns of economic production and consumption promoted by economic globalisation. For those that can pay, the globalised economy can provide wasteful lifestyles, which increasingly depend on large volumes of water and electricity. For the poorer strata of society, however, globalisation has brought new threats to their livelihood and additional pressures on shared natural resources. The consequence is that the ongoing adjustments continue to stir conflicts and provoke bitter reactions among poorer citizens and environmentalists across the region.

By and large, the search for more flexible water policies in Latin American countries has been contained by their technocratic insistence on the internalisation of costs and the optimisation of resources, while social justice and collective responsibilities for the degradation of shared water are left out of the equation. The priority of economic rationality for the solution of water problems only sustains a regime of environmental exploitation and social exclusion that historically has characterised water management in the continent. It has been mentioned elsewhere that market solutions are inadequate to deal with stochastic and complex ecological systems, because it creates a 'policy lock-in' that precludes genuine alternatives to the hegemonic statehood model (Bowles, 2004). In other words, the priority given to the economic dimension of water management is nothing but the mainstream political paradigm reflecting its view of itself. In addition, Bowles (2004) observes that market forces have more than a merely allocative role, but also exert a disciplinary function that in reality operates through the asymmetric use of power. At the same time, while acknowledging the harmful impacts of market pressures, not all the problems of the new water regime are solely the result of economic priorities stimulated by the state. On the contrary, there are other fundamental factors that contribute locally to management failures. It is precisely the powerful articulation between the hegemony of market-based regulation and local power asymmetries that have ultimately been responsible for the insufficient results of the water reforms in Latin America.

The last pages of this chapter have demonstrated that the search for flexible environmental statehood in Brazil and in Latin America at large entailed the adoption of market-based instruments, including utility privatisation, water user charges and the payment for ecosystem services, as the supposed response to old and new problems. Nonetheless, the environmental results of the new water regulation have been, at best, disappointing. The new regime has both aggravated stakeholder conflicts and legitimised the negative impacts of more intensive water users via operation licences and bulk water charges. It means that, in practice, little has changed: the stability of water systems and the fundamental rights of deprived social groups continue to be forfeited to the need for more dams or the exploitation of catchment resources. Even in catchments with more meaningful public mobilisation and solid structures of public representation, the degrading trends remain unaffected (paradigmatic examples in Brazil are the river basins of Sinos, São Francisco, Piracicaba and Paraíba do Sul, among others). This apparent paradox of novel legal approaches that reproduce old problems can be explained by the bureaucratisation of relations between society and the rest of socionature under hegemonic market-based policies. The persistence of water management problems is certainly acknowledged by many scholars, however there is still limited scrutiny of the systematic failures of the new regulatory regime. It is rare to see authors willing to investigate why technological improvements, public participation and mitigation measures have been systematically sidestepped by the accumulation strategies of present-day capitalism.

As in other countries undertaking similar institutional reforms, the economic agenda has largely underpinned the reform of the water sector and, ultimately, the introduction of flexible environmental statehood in Brazil. An intriguing example is provided by the 2006 National Water Plan, which explicitly claims that economic growth is a precondition for the solution of water-related problems. The plan described three future scenarios for water sector: 'water for all', 'water for some' and 'water for few'. The main difference between those scenarios was the projected annual rate of GDP growth, respectively 4.5 per cent, 3.5 per cent and 1.5 per cent per year (MMA, 2006). As can be seen from the different rates, according to these scenarios, water access would be universalised in Brazil only with a higher rate of economic growth, whilst environmental restoration depends on the good performance of the economy. Here, as in other documents, the association between water management and economic growth ultimately instils a particular pattern of social relations that are fraught with tensions and contradictions. If, in the past, the emphasis was on the construction of engineering works, the current water management reforms aim to remove obstacles to economic production (e.g. river pollution and water scarcity), at the same time as they raise new alternatives for capital accumulation (e.g. environmental consultancy and the payment for ecosystem services). In the same way as economic agents need to invest in technology to increase relative surplus value and also contain the workforce, there is a need to remove ecological degradation in order to restore accumulation conditions and contain the threats to the stability of economic systems. Before market-based responses can be adopted, it is essential that the monetary value of water be quantified and discursively normalised, which is achieved by the imposition of water charges (i.e. the sophisticated approaches developed by mainstream economists to estimate water charges have resulted in the institutionalisation of a common monetary basis among water users). The powerful symbolism of the monetary value of water enables the reinsertion of degraded environmental systems into production relations via the commodification of restoration and conservation measures.

In the end, the Kantian-informed transition to more flexible environmental statehood and regulation in Brazil served primarily the stronger interest groups that sponsored the institutional reorganisation, and not the wider socioecological demands. The consequence is the persistence of disputes, conflicts and degradation (not to mention repeated cases of corruption and waste of resources) in marked contrast to the lip service paid to public participation. The failures and insufficiencies of the initial experiences of statehood flexibilisation, as illustrated by the Brazilian reform of the water sector, mean that a more comprehensive, pro-commons framework of environmental statehood is still needed. The problems of the transition to flexible environmental statehood will be examined in more detail in the next chapter. After that, the following chapter contains the main argument put forward in this book: that such recent adjustments have represented the fulfilment of the Hegelian plans for the state system.

Notes

1 Moreover, one of the political prices paid by the Reagan administration was the questionable legitimacy of those same changes, which was only resolved a few years later (under President Clinton) through a new round of regulatory adjustments aimed at securing higher levels of justification and public support.
2 Permissive law, in the sense suggested by Kant, is that which applies to actions which are neither obligatory nor prohibited and that people are free to submit to as they please.
3 Public irrigation schemes have also been increasingly transferred to private enterprise, normally through a fixed-term concession of land and infrastructure.
4 However, since 2012 there are signs that utility privatisation may be returning to the policy priorities of the federal government and many state (provincial) administrations.
5 FGTS stands for Severance Pay Fund, which provides severance pay in cases of unjustified dismissal. It creates a savings fund for workers and also to finance housing and water and sanitation programmes.

Bibliography

Abers, R.N. and Keck, M.E. 2006. Muddy Waters: The Political Construction of Deliberative River Basin Governance in Brazil. *International Journal of Urban and Regional Research*, 30(3), 601–22.
Andrews, R.N.L. 1984. Deregulation: The Failure at EPA. In: *Environmental Policy in the 1980s: Reagan's New Agenda*, Vig, N.J. and Kraft, M.E. (eds). CQ Press: Washington D.C., pp. 160–80.
Anuatti-Neto, F., Barossi-Filho, M., Carvalho, A. and Macedo, R. 2003. *Costs and Benefits of Privatization: Evidence from Brazil*. IADB: Washington, DC.
Appadurai, A. 1986. Introduction: Commodities and the politics of value. In: *The Social Life of Things: Commodities in Cultural Perspective*, Appadurai, A. (ed.). Cambridge University Press: Cambridge, pp. 3–64.
Appadurai, A. 2005. Materiality in the Future of Anthropology. In: *Commodification: Things, Agency, and Identities*, Van Binsbergen, W. and Geschiere, P. (eds). Lit: Münster, pp. 55–62.
Bowles, S. 2004. *Microeconomics: Behavior, Institutions and Evolution*. Russell Sage Foundation: New York.
Brand, U. and Görg, C. 2008. Post-Fordist Governance of Nature: The Internationalization of the State and the Case of Genetic Resources – a Neo-Poulantzian Perspective. *Review of International Political Economy*, 15(2), 567–89.
Breitenbach, A. 2005. Kant Goes Fishing: Kant and the Right to Property in Environmental Resources. *Studies in History and Philosophy of Science Part C: Studies in History and Philosophy of Biological and Biomedical Sciences*, 36(3), 488–512.
Brenner, N., Peck, J. and Theodore, N. 2010. Variegated Neoliberalization: Geographies, Modalities, Pathways. *Global Networks*, 10(2), 182–222.
Brenton, T. 1994. *The Greening of Machiavelli: The Evolution of International Environmental Politics*. Earthscan: London.
Britto, A.L.P. and Silva, R.T. 2006. Water Management in the Cities of Brazil: Conflicts and New Opportunities in Regulation. In: *Urban Water Conflicts: An Analysis of the Origins and Nature of Water-related Unrest and Conflicts in the Urban Context*. IHP/UNESCO: Paris, pp. 39–51.
Coppetec. 2006. *Plano de Recursos Hídricos da bacia do Rio Paraíba do Sul*. AGEVAP: Resende.

Deleuze, G. 2008. *Kant's Critical Philosophy*. Trans. H. Tomlinson and B. Habberjam. Continuum: London and New York.

Ellis, E. 2005. *Kant's Politics*. Yale University Press: New Haven and London.

European Commission. 1989. *Proposal for a Council Regulation (EEC) on the establishment of the European Environment Agency and the European Environment Monitoring and Information Network*. COM(89) 303 final.

European Commission. 1993. *Towards Sustainability, a Policy and Strategy for the Environment and Sustainable Development within the European Community*. Fifth EC Environment Action Programme (1993–2000). Official Journal of the European Communities in C 138 of 17 May 1993.

European Commission. 2001. *White Paper on European Governance*. COM (2001) 428 final.

Fearnside, P.M. 1997. Environmental Services as a Strategy for Sustainable Development in Rural Amazonia. *Ecological Economics*, 20(1), 53–70.

Féres, J., Thomas, A., Reynaud, A. and Serôa da Motta, R. 2005. *Demanda por Água e Custo de Controle da Poluição Hídrica nas Indústrias da Bacia do Rio Paraíba do Sul*. Document No. 1084. IPEA: Rio de Janeiro.

Formiga-Johnsson, R.M., Kumler, L. and Lemos, M.C. 2007. The Politics of Bulk Water Pricing in Brazil: Lessons from the Paraíba do Sul Basin. *Water Policy*, 9(1), 87–104.

Guyer, P. 2006. *Kant*. Routledge: London and New York.

Hall, D. and Lobina, E. 2007. Profitability and the Poor: Corporate Strategies, Innovation and Sustainability. *Geoforum*, 38(5), 772–85.

Harvey, D. 2005. *A Brief History of Neoliberalism*. Oxford University Press: Oxford.

Heller, P. 2001. Moving the State: The Politics of Democratic Decentralization in Kerala, South Africa, and Porto Alegre. *Politics and Society*, 29(1), 131–63.

Hildebrand, P.M. 2002. The European Community's Environmental Policy, 1957 to '1992': From Incidental Measures to and International Regime. In: *Environmental Policy in the European Union*, Jordan, A. (ed.). Earthscan: London, pp. 13–36.

Holland, A.-C.S. 2005. *The Water Business: Corporation versus People*. Zed Books: London and New York.

Hurrell, A. 1994. A Crisis of Ecological Viability? Global Environmental Change and the Nation State, *Political Studies*, 42(supplement s1), 146–65.

IBGE. 2004. *Atlas de Saneamento*. IBGE: Rio de Janeiro.

Ioris, A.A.R. 2007. The Troubled Waters of Brazil: Nature Commodification and Social Exclusion. *Capitalism Nature Socialism*, 18(1), 28–50.

Ioris, A.A.R. 2009. Water Reforms in Brazil: Opportunities and Constraints. *Journal of Environmental Planning and Management*, 52(6), 813–32.

Jordan, A., Wurzel, R.K.W. and Zito, A.R. (eds). 2003. *'New' Instruments of Environmental Governance: National Experiences and Prospects*. Frank Class: London and Portland, OR.

Kant, I. 1929 [1781]. *Immanuel Kant's Critique of Pure Reason*. Trans. N.K. Smith. Macmillan: Basingstoke and London.

Kant, I. 1991. *Kant's Political Writings*. Ed. H.S. Reiss and trans. H.B. Nisbet. Cambridge University Press: Cambridge.

Kant, I. 1996 [1797]. *The Metaphysics of Morals*. Ed. and trans. M. Gregor. Cambridge University Press: Cambridge.

Kant, I. 2002 [1785]. *Groundwork for the Metaphysics of Morals*. Ed. T.E. Hill and A. Zweig; trans. A. Zweig. Oxford University Press: Oxford.

Kaufman, A. 1999. *Welfare in the Kantian State*. Oxford University Press: Oxford.

Kersting, W. 1992. Politics, Freedom, and Order: Kant's Political Philosophy. In: *The Cambridge Companion to Kant*, Guyer, P. (ed.). Cambridge University Press: Cambridge, pp. 342–66.

Kosoy, N., Martinez-Tuna, M., Muradian, R. and Martinez-Alier, J. 2007. Payments for Environmental Services in Watersheds: Insights from a Comparative Study of three cases in Central America. *Ecological Economics*, 61(2–3), 446–55.

Lefebvre, H. 1972. *The Sociology of Marx*. Trans. N. Guterman. Penguin Books: Harmondsworth.

Lefebvre, H. 2009. *State, Space, World: Selected Essays*. Trans. Moore, G., Brenner, N. and Elden, S. University of Minnesota Press: Minneapolis, MN.

Liefferink, D. 1996. *Environment and the Nation State*. Manchester University Press: Manchester and New York.

Mansfield, B. 2007. Articulation between Neoliberal and State-oriented Environmental Regulation: Fisheries Privatization and Endangered Species Protection. *Environment and Planning A*, 39(8), 1926–42.

Marx, K. 1973 [1857–8]. *Grundrisse*. Trans. M. Nicolaus. Penguin and New Left Review: London.

MMA. 2006. *Plano Nacional de Recursos Hídricos*. Ministry of the Environment: Brasília.

Mulreany, J.P., Calikoglu, S., Ruiz, S. and Sapsin, J.W. 2006. Water Privatization and Public Health in Latin America. *Revista Panamericana de Salud Pública*, 19(1), 23–32.

OECD. 2011. *Water Governance in OECD Countries: A Multi-Level Approach*. OECD Studies on Water. OECD Publishing.

O'Hagan, T. 1987. On Hegel's Critique of Kant's Moral and Political Philosophy. In: *Hegel's Critique of Kant*, Priest, S. (ed.). Oxford University Press: Oxford, pp. 135–59.

Oliveira Filho, A. 2006. Institucionalização e Desafios da Política Nacional de Saneamento: Um balanço. *Proposta*, 110, 12–23.

Paterson, M., Doran, P. and Barry, J. 2006. Green Theory. In: *The State: Theories and Issues*, Hay, C., Lister, M. and Marsh, D. (eds). Palgrave Macmillan: New York, pp. 135–54.

Ribeiro, E.M., Galizoni, F.M., Calixto, J.S., Assis, T.P., Ayres, E.B. and Silvestre, L.H. 2005. Gestão, Uso e Conservação de Recursos Naturais em Comunidades Rurais do Alto Jequetinhonha. *Estudos Urbanos e Regionais*, 7(2), 77–99.

Robertson, M. 2007. Discovering Price in all the Wrong places: The Work of Commodity Definition and Price under Neoliberal Environmental Policy. *Antipode*, 39(3), 500–26.

Saner, H. 1973. *Kant's Political Thought: Its Origins and Development*. Trans. E.B. Ashton. University of Chicago Press: Chicago and London.

Vargas, M. 2005. *O Negócio da Água – Riscos e Oportunidades das Concessões de Saneamento à Iniciativa Privada: Estudos de Caso no Sudeste Brasileiro*. Annablume: São Paulo.

Vosti, S.A., Braz, E.M., Carpentier, C.L., d'Oliveira, M.V.N. and Witcover, J. 2003. Rights to Forest Products, Deforestation and Smallholder Income: Evidence from the Western Brazilian Amazon. *World Development*, 31(11), 1889–901.

While, A., Jonas, A.E.G. and Gibbs, D. 2010. From Sustainable Development to Carbon Control: Eco-State Restructuring and the Politics of Urban and Regional Development. *Transactions of the Institute of British Geographers*, NS 35, 76–93.

Williams, H. 2003. *Kant's Critique of Hobbes*. University of Wales Press: Cardiff.

Wunder, S. 2007. The Efficiency of Payments for Environmental Services in Tropical Conservation. *Conservation Biology*, 21(1), 48–58.

Wurzel, R.K.W. 2008. European Union Environmental Policy and Natura 2000. In: *Legitimacy in European Nature Conservation Policy: Case Studies in Multilevel Governance*, Keulartz, J. and Leistra, G. (eds). Springer, pp. 259–82.

5 The contingent relation between flexible environmental statehood and the neoliberalisation of socionature

The increasing neoliberalisation of socionature

The previous chapter discussed the transition, at the end of the twentieth century, to a new model of environmental statehood characterised by more flexible approaches for dealing with socioecological problems. After several decades under the sphere of influence of conventional statehood, the institutional framework started to shift towards more comprehensive and less punitive alternatives. At the same time, it was necessary, even if at a rhetorical level, to offer compensation for the erosion of the commons and engage wider society in a sense of collective responsibility. This meant that, because of the complexity of contemporary environmental matters, there was a growing acknowledgement by politicians and policy-makers that some concessions had to be made to wider society and to groups directly affected by environmental degradation. The transition to flexible environmental statehood had the double purpose of enhancing the response to environmental concerns and lubricating the mechanisms of economic production and capital accumulation. The reform of environmental statehood ultimately represents the most recent chapter in the long process of bringing the commons into mainstream socioeconomic activities. It was left to the state, as a socioecological relation, to play the most crucial role in the subtle administration of anti-commons pressures and the appeasement of public opinion. Those in charge of the state had to learn how to manage the changing politico-ecological circumstances and, in the process, adjust the configuration and functioning of the state.

At the centre of this institutional transition was the Kantian defence of rational thinking and high levels of morality, which should operate within the sphere of private property and a liberal, but still strong, state apparatus. The external image of such transformations was provided by the language of environmental governance and the associated concepts of the green economy, ecological modernisation and sustainable development. As stated by Bernstein (2001: 214), the "growing importance and prominence of environmental concerns in global governance owes much to its formulation in norms of liberal environmentalism". Kant's dualistic ontology (noumenal–phenomenal) was particularly consistent with the rhetorical calls for sustainable development. A sustainable world, as vaguely defined in official policies, is essentially noumenal and distant from a phenomenological reality that actually moves away from sustainability. For Kant

(1929), the reality of the world can only be apprehended with methods and ideas that are established a priori, which was instrumental in demonstrating that environmental sustainability could be achieved through the introduction of new technologies and management approaches (and, obviously, without altering mainstream socioeconomic trends). Likewise, the adoption of policies and procedures related to the search for sustainable development evolved on a par with the expansion of a mass consumption and mass waste economy. Kant's system of moral conduct – based on abstract principles and universal rules of morality – was also highly effective at disguising the contradiction between growing economic freedoms and weakening socioecological guarantees. In the process of establishing flexible environmental statehood, rights and freedom were built from the perspective of the individual and not necessarily related to the collective ownership and protection of the dwindling commons. Equally, the root causes of environmental problems remained practically unquestioned (i.e. anti-commons priorities, intensified exploitation of resources, double degradation of society and the rest of nature, etc.), given that the prime objectives of the renovation of environmental statehood were not the resolution of shared environmental problems but the containment of political reactions and the creation of new avenues for private capital accumulation. In the end, the period since the 1980s has witnessed multiple procedures aimed at renewing ('enlightening', in the Kantian sense) environmental statehood while preserving and facilitating the fulfilment of hegemonic ambitions.

One specific ramification of the search for a more flexible environmental statehood since then has been the implementation of a number of procedures that together lead to the growing *neoliberalisation of socionature*. The neoliberalisation of socionature has in effect become a central element of the late spread of Western modernity and the consolidation of a global-market society. As typically happens in any new area of capitalist expansion, the neoliberalisation of socionature triggered new demands for rationalisation and political legitimation. Bernstein (2001: 214) properly observes that "the legitimation of environmental concerns in the international political economy has involved a process of introducing ideas about the environment that, to gain legitimacy, required some compatibility with" the dominant economic order. This includes changes in both private and public spheres of life according to individualist values and short-term economic priorities. In terms of the consolidation of flexible environmental statehood, Kantianism was no longer enough, and the political ideas of another major European philosopher, Hegel, whose elaboration on the legitimacy of the bourgeois state proved to be more dynamic and fertile than the more abstract system proposed by Kant, were required. The influence of Hegel will be analysed elsewhere in this book. For the moment, the current chapter aims to explain the intricacies and shortcomings of the contingent association between the neoliberalisation of socionature and the pursuit of new bases of environmental statehood.

The advance of neoliberalism obviously did not remove the role of the state as the key mediator and promoter of socionatural interactions. On the contrary, it was precisely the state which continued to be the primary agent of neoliberalism, maintaining its central role in governing socionatural relations. Socionature neo-liberalisation went beyond the reorganisation of the state apparatus, at the same time as the state played a central role in the preservation and promotion of liber-alising strategies. The promotion of neoliberal agendas has largely depended on the executive procedures previously introduced by the Keynesian state, which had to adapt itself in order to guarantee the viability of new market-based pol-icies. The result was the unfolding of two interrelated processes: first, a series of changes in the 'public' sphere of the state and in the 'private' sphere of eco-nomic and social life, and, second, transformations in the internal coordination of government strategies and the complex interconnections through networks (Picciotto, 2011). These processes have major consequences for environmental statehood and the organisation of its associated state-fix. On the one hand, the neoliberal state has had to react to pressing demands to resolve environmental degradation and related conflicts, which has required some level of state inde-pendence from the groups involved in disputes. On the other hand, the advance of neoliberalism by the state has been an integral driving-force behind the rein-vigoration of capitalist social relations, which makes it permeable to hegemonic political interests and undermines its ability to contain the environmental degradation caused by capital accumulation pressures.

Before going any further, three main comments are required to situate the analysis of socionature neoliberalisation in relation to environmental statehood. First, neoliberalisation is a phenomenon that began with the ideological "separa-tion of nature and society and then reconnected them by reductively constructing 'nature' so that it can be encompassed within 'economy'" (McAfee and Shapiro, 2010: 581). This happens through heterogeneous and spatio-temporally differen-tiated processes, such as privatisation, enclosure of the commons, market-proxies and the monetary valuation of ecosystems (Heynen and Robbins, 2005). It has entailed a complex set of new procedures and discursive constructions that not only preserved the foundations of the capitalist economy, but incorporated factors previously regarded as 'extra-economic' into the global accumulation of capital (e.g. ecosystem services, atmospheric composition, environmental con-servation, etc.). As previously observed by Marx (1973: 531), the privatisation of public services and their migration into the domain of the works undertaken by capital itself "indicates the degree to which the real community has consti-tuted itself in the form of capital." Through various monetising strategies, neo-liberal environmental policy-making started to place growing importance on economic instruments of environmental management, such as taxes, subsidies and tradable emission permits. Even in the absence of explicit market mecha-nisms for dealing with natural resources, neoliberal policies have made use of 'market proxies' as efficiency-seeking incentives (e.g. in the case of regulatory charges related to the economic value of resources or ecosystems).

Second, it should be noted that in recent years the literature on the neoliberal-isation of socionature has steadily grown and incorporated a diversity of investi-gative approaches. It is clear from these texts that socionature neoliberalisation is a process that evolves through a complex mixture of ideological discourses and political pressures that interconnect mechanisms of co-optation and subordination. The neoliberalisation of nature has typically been described in the academic literature as an intricate phenomenon that combines market pressures and the re-regulation of environmental management, however the synergies between the socioecological and politico-economic dimensions, and the shared agency between nature and society, have rarely received sufficient consideration (Bakker, 2010). Despite the growing body of scholarly work, there still remains a significant inconsistency among studies of nature neoliberalisation due to the recurrent intention to accommodate dissimilar, highly idiosyncratic experiences under the same analytical categories (Castree, 2008a). In the case of those studies, the search for similarities can easily become a formal rather than a substantive exercise. In effect, the basis and the practice of neoliberalisation are dynamic and elastic, which is what is required to operate in a world that is 'more-than-neoliberal' (Bakker, 2010). Another major gap in the existing literature on the neoliberalisation of socionature is the fact that a more dynamic examination of the role of the state is still largely missing. So far most of the scholarly attention has been on the reformatting of environment–society relations, but much less consideration has been given to the adjustment of neoliberal policies and to the reconfiguration of the state apparatus to deal with socioecological barriers. Examples of these barriers are ecosystem processes not easily inserted in market-like transactions, such as biodiversity conservation and water quality, or the opposition of local communities affected by the privatisation of utilities and by the increased exploration of resources and territories.

The third initial observation is that the new approaches and rationalities associated with the neoliberalisation of socionature were not politically neutral, but necessarily emerged from the political commitments of the state apparatus and the overall balance of political power. The neoliberal state is commonly designed as an idealised formation that imitates business management, favours market institutions and should serve primarily the interests of private capital. Environmental statehood under neoliberalising strategies preserves core capitalist objectives and long-established property relations. Those objectives can only be pursued through the incorporation of more features of the nonhuman world into market transactions, the privatisation of hitherto protected areas of socionatural interaction and the intensification of socionature commercialisation. Instead of simply deregulating the market, as advocated by neoliberal authors, the advance of neoliberalism has happened through the re-regulation of socioeconomic and socioecological processes. Another central aspect of the neoliberal reconfiguration of the state apparatus is its ability to profit, in terms of its institutional renovation, from new socioeconomic opportunities. The environmental demands imposed on the neoliberal state have gradually resulted in a creative, innovative process of statecraft, which includes renewed mechanisms of socionatural interaction and the persistent search for political legitimation.

In addition to these three preliminary observations that underline the need for more detailed assessments, it should be noted that in practice the neoliberalisation of socionature has produced only limited responses to mounting environmental problems. It is clear that the international experience of socionature neoliberalisation has proved highly problematic even for the purpose of capital accumulation. Probably the most compelling example of its limited achievements is the failure to curb anthropogenic climatic change with the introduction of carbon markets and renewable energy subsidies. A serious shortcoming of those approaches, which mirrors the wider contradiction of capitalist relations of production, has been the falling rate of profit from the commodification of socionature. By resorting to Marxian theory, it is possible to verify that the organic composition of capital (i.e. related to the fixed capital invested in production) tends to rise over time because of the adoption of increasingly costly methods and techniques of production. After an initial easy appropriation of natural resources and ecosystem features, profitability declines because of both additional geographical distances and the complexity of the socionatural features to be commodified. Profitability is also normally affected by the reaction of local communities, in alliance with groups in other locations, to the expansion of capitalist relations of production and the associated loss of their livelihoods. Moreover, for Marx, tendencies interact with counter-tendencies and, in practice, the contradictory basis of capitalist exploitation does not necessarily lead to an automatic crisis. Such "Marxian laws express the key material forces constituted by capitalist social relations, what Marx calls tendencies" (Fine and Saad-Filho, 2010: 95). These are all historic-geographical processes that evolve according to an unequal balance between production and exchange of goods and services. The economic and administrative results of socionature neoliberalisation directly depend on the concrete properties of the socioecological systems that are being incorporated into market-like transactions.

Still, when considering the complex landscape of the neoliberalisation of socionature, most authors have so far placed great emphasis on institutional changes and business-like transactions, but not enough on the intricate basis and constantly evolving practices of neoliberalisation. From a politico-ecological perspective, a specific problem is the proliferation of framing concepts in the attempt to condense in a few rules the remarkable complexity of the neoliberalisation of socionature. We can refer to the literature on the neoliberalisation of water management to illustrate the difficulty with the terminology. For instance, Bakker (2005) describes the neoliberalisation of water as the product of three overlapping forces, namely commodification (market exchange of water processes previously outside the sphere of the market), commercialisation (adoption of commercial principles and methods) and privatisation (changes in resource and utility ownership). Smith (2004) differentiates between privatisation (transfer of ownership) and corporatisation (service delivery in which the state retains control but also delegates the management of public water services to private firms). Budds and McGranahan (2003) consider that privatisation is a generic term that can be used also in reference to private sector investment (e.g. BOTs) and public–private partnerships (PPPs), therefore blurring the boundaries between privatisation, corporatisation and commercialisation. The main limitation of these types of classification is not a focus only on the procedural aspects of the advance of neoliberalism over nature, but the reluctance to explain the more strategic and adaptive attributes that have secured the persistence, often in disguised ways, of neoliberalising agendas.

The idiosyncratic manifestations of those typologies call for a more comprehensive analytical framework able to account for the elasticity and heuristic properties of nature neoliberalisation. Instead of concentrating on specific drivers and framing concepts, it is necessary to address the totality of the relations between society, nature and the state under the neoliberal waves of accumulation. In that context, the neoliberalisation of water can serve as an evocative example of the wider neoliberalisation of socionature, without forgetting that it is more than the simple advance of commodifying mechanisms and market-friendly procedures, but that it has also entailed a series of interrelated institutional and political changes. Technical, economic and political adjustments are central attributes of water neoliberalisation and the fluid interplay between these three dimensions explains the insertion of water use and conservation into market-based transactions. By focusing on these three interrelated dimensions, it is possible to understand the actual evolution of water neoliberalisation as a succession of strategies that lead to different institutional changes according to specific demands and constraints. This means that the differences between the processes of water neoliberalisation occurring in countries, cities and locations around the world derive from the specific permutation between these three dimensions. This conceptual framework also facilitates a fresh look into political reaction and popular resistance to the imposition of neoliberalised approaches on nature and society.

In general terms, it can be said that the experience of water regulatory reforms in the last three decades has been marked by a succession of moments when one dimension is apparently more evident than the others. The three dimensions are always present in any process of water neoliberalisation, but one dimension seems to prevail in a given historical period according to concrete politico-geographical circumstances. The first phase of the neoliberalisation of water, already occurring in the late 1980s, was primarily concerned with reducing the negative externalities generated during the welfarian, developmentalist period. Previous assessments of the neoliberalisation of water have overlooked the fact that the agenda of reforms initially focused on paving the way for the subsequent adoption of market-friendly management solutions. In that sense, the introduction of new water policies and technical adjustments represented an early, but necessary, stage of the flexibilisation of water management and the creation of a greater space for non-governmental players (private companies, but also NGOS and civil society organisations). This first phase of water neoliberalisation was, therefore, centred on a *techno-environmental* dimension and encompassed a set of measures designed to cope with operational inefficiencies and environmental impacts caused during the Keynesian phase. Some of the initial evidence of the new direction of water management policies was the recommendations adopted at the United Nations Water Conference in Mar del Plata, Argentina, in 1977, which were later reinforced during the UN International Drinking Water Supply and Sanitation Decade between 1981 and 1990. The official documents of the Conference introduced pivotal concepts that gradually altered water management in the subsequent years (i.e. calls for holistic catchment assessments, pollution prevention, water balance models, public participation in decision-making, and improvements in monitoring and data management). Most countries responded by passing legislation that stipulated Environmental Impact Assessments and similar regulatory requirements. In particular, the doctrine of integrated water resources management (IWRM) became a crucial influence on new policies and legislation.

Efforts in terms of techno-environmental improvements continued to inspire the formulation of public policies, but it was not until the 1990s that the neoliberalisation of water began to move into arenas formerly inaccessible to market-like transactions (Sangameswaran, 2009). The emphasis shifted from technical matters to a more uncompromising economic rationality as an attempt to forge novel business opportunities in the water industry. The language changed from a 'focus on mitigating environmental impacts' into the internalisation of 'environmental externalities' and the removal of 'state failures'. This was the tone of the 1992 Dublin Statement on Water and Sustainable Development, which asserted that water is a finite and vulnerable resource that has an economic value and should be recognised as an economic good. In a period when the state was blamed for the breakdown of water services, the so-called 'Dublin Principles' offered a logical, and also very expedient, justification for further market-based relations around water, which chiefly utilised a shift from water supply towards demand management and higher levels of economic and technical efficiency. This more explicit economic dimension of water management corresponds to the *monetisation* of water, that is, the attachment of monetary value to water and the handling of water as an economic resource amenable to market transactions and 'rational choice' theory. The monetisation of water presupposes the standardisation of socioecological values through the use of monetary standards (as the implicit reference of value), which is achieved primarily via the introduction of water permits and charges, as well as through the language of cost-benefit analysis underpinning responses to water problems. The monetary valuation of water played a key normative role in terms of environmental governance because it could provide the markets with the information needed to pursue commercial-like relations.

Both techno-environmental adjustments and the monetisation of water were strongly encouraged by multilateral agencies, such as the World Bank and the World Water Council. It was particularly through structural adjustment plans that the privatisation of water utilities and the adoption of market-friendly regulation were implemented around the world. At the same time, the spreading of water neoliberalisation was lubricated by hundreds of international summits (such as the World Water Forum, in The Hague, in 2000). Nonetheless, the insistence on the role of the private sector in water supply and environmental management also faced growing resistance in the Global South. Disastrous cases of utility privatisation – as in Bolivia, Argentina and South Africa, where operation contracts were seen to dishonestly favour multinational corporations bringing about higher tariffs, corruption and the cancellation of service to many of the poorest households – prompted a worldwide reaction against the ideology of water neoliberalisation. There was a growing gap between the promises of the private participation model (advocated mainly by the World Bank) and the reality of the water services in the countries that received loans and collaborative funds. Critical problems included the failure to find additional financial sources, maintain the quality of operation and expand household connections. General acceptance of the orthodox neoliberal formula became gradually more difficult to maintain and, as a result, the agenda of water neoliberalisation was forced to incorporate concepts such as adaptive management, transition management and multi-stakeholder participation. Recognising that the early neoliberal policies were not particularly intended to serve low-income water users, at the end of the 1990s a pro-poor rhetoric emerged as a response to increasing political unrest. In addition, the United Nations declared the period 2005–15 as the 'International Decade for Action – Water for Life', which emphasised further water management issues such as the importance of culture, race and gender. Therefore, the third moment of water neoliberalisation (particularly since the Kyoto Forum in 2003) was centred on the *legitimisation* of new regulation and service provision, but without any significant departure from the ultimate aims of capital accumulation through environmental management and service provision.

It is important to recognise that the above scheme of the evolution of water neoliberalisation did not happen in exactly the same way, or in the same order, in all countries, catchments and locations. In effect, the manifestation of each dimension of water neoliberalisation did not happen simultaneously around the world, but was obliged to follow specific socioecological and political circumstances. The outcome of the neoliberalisation of water has depended on the tension between capital accumulation goals and the extra-economic reactions offered by social and ecological systems. In many cases, because of political opposition, the legitimisation effort has had to come together with economic and techno-environmental adjustments. In other cases, the emphasis may have been restricted to the techno-environmental improvements that were needed by other economic sectors, rather than the creation of new avenues of accumulation directly related to water management. Despite the idiosyncratic features of individual experiences, there are always important synergies between the three dimensions, which define the level of success of water neoliberalisation programs. For instance, techno-environmental initiatives are required to make water more easily available to monetised relations. Similarly, the acceptability of market-like responses requires the production of tangible techno-environmental results, which together help to reinforce political legitimisation.

The three dimensions included in this analytical framework show the vitality of the neoliberalisation of socionature and the ability of the capitalist state to adapt and move from a welfare-developmentalist emphasis to neoliberal, market-based state strategies. The neoliberalisation of socionature is only the most recent stage of the overall 'capitalisation of nature', that is, the appropriation of the natural world by capitalist forces and relations of production according to unequal property relations and the drive for capital accumulation. The main advantage of the framework suggested in this book is to clearly demonstrate the dynamic and contingent evolution of the neoliberalisation of water as a multidimensional process that moves forward according to particular historico-geographical circumstances. While for some authors the neoliberalisation of socionature is narrowly associated with the monetisation dimension, according to the framework suggested, the divestiture of public utilities constitutes only one moment of the overall neoliberalising policies. Instead of focusing on one specific framing category, the proposed conceptual model captures the necessary interplay between the structuring dimensions. In this way, it accounts for the appropriation of public engagement and environmental restoration by policies aimed at developing new routes of capital accumulation. It also helps to elucidate why a significant number of critical reactions to the neoliberalisation of nature have been systematically absorbed by the official mechanisms of governance and conflict resolution. Furthermore, the framework facilitates the perception of the totality of neoliberalisation and the specific alterations needed to sustain the process of institutional reforms.

The synergies between the internal dimensions of the neoliberalisation of socionature, with special reference to the transformations in the water sector, will be demonstrated in the next two sections with two examples of reforms in the state apparatus directly related to the implementation of more flexible models of environmental statehood. The first will be a discussion of the adjustments in the European Union associated with the new water directive, with a focus on the Scottish experience, and the second will be an examination of the reform of the water services of Lima, Peru. In both cases, there is a clear transformation of the symbolic and material basis of the use and conservation of water, which coincides with renewed opportunities for the involvement of the private sector in environmental management and with novel attempts to communicate with and engage the general public (although firmly within the existing economic and sociopolitical tendencies).

European Union's framework of environmental statehood

Environmental statehood, as defined in Chapter 2, consists of a set of ideas and measures developed by the state, in specific historical and geographical circumstances, to restrain the access, use and abuse of territorialised natural resources and ecosystem services. This involves the control of the handling of private assets (e.g. land, industries, mineral reserves, etc.) according to what is considered to be the public interest and the long-term needs of the economy. The implementation of environmental statehood closely follows the requirements of the wider capitalist order, based on the perpetuation of social inequalities, the private appropriation of commons and the double exploitation of society and the rest of nature. Because of its biased orientation, the enforcement of environmental regulation is typically associated with scepticism, political reactions and social tensions. The articulation between public and private dimensions of environmental management and the mechanisms of state control over socionatural phenomena are intrinsically prone to contestation. The very definition and interpretation of what is meant by the boundaries between society and nature is something inherently political and reflects mechanisms of social inclusion and exclusion.

One of the most paradigmatic demonstrations of the connection between state ideologies, political clashes and the appropriation of environmental agendas by hegemonic economic sectors has been the organisation of environmental statehood in the realm of the European Union (EU). Its environmental legislation is seen as standing out "as a notable European and international policy achievement, when compared with other EU issue areas" (Zito, 2000: 2). Nonetheless, the EU public sector is anything but simple. It extends from local authorities to the administration of member states and, eventually, the interstate structure (i.e. the European Commission, Parliament, Council of Ministers and Court of Justice), which have to work together in the approval and implementation of environmental policy, legislation and regulation. This has required a dynamic process of statecraft at the European, national and sub-national levels of public administration, with constant amendments and additions of policy units and regulatory agencies. The responses formulated by the EU had to be developed within the political space available and according to economic and sociopolitical priorities.

The more recent changes in the structure and operations of the EU state apparatus have corresponded to a transition from its initial rigid rules, deterrence strategies and undifferentiated treatment of regulated actors (particularly in the 1970s) to a new wave of environmental regulation (since the 1990s) which emphasises continuous improvement according to environmental indicators, rules backed by sanctions to influence behaviour and a more nuanced understanding of factors that affect the performance of regulated social actors (Fiorino, 2006). It means a shift from command-and-control mechanisms of regulation (characteristic of the welfare state and inspired in the notions of authority and law enforcement put forward by Hobbes) to the more flexible model associated with the search for sustainability, governance and adaptive management (the hallmark of the neoliberalism of socionature influenced by the political theories of Kant and, especially, Hegel).

In the first decades of the Union, some initial pieces of environmental legislation were already being introduced and gradually formal environmental controls started to occupy a more central position in European negotiations. By the end of the 1980s, legislation became increasingly more complex, due to the growing number of countries and issues involved. The environmental question was an important element of negotiation in the run up to the Maastricht Treaty of 1992, which formally created the European Union. All EU member states are expected to comply with an extensive body of environmental legislation, which has repercussions on commercial transactions and trade agreements with the rest of the world. The conventional model of environmental statehood, adopted in the 1960s and 1970s, was criticised for being too onerous, heavy-handed and allowing insufficient space for public participation, which triggered a gradual replacement and the introduction of plans and regulatory efforts from a more flexible and market-friendly perspective. The new environmental statehood was adopted to stimulate technological innovation, preserve economic competitiveness and avoid unnecessary burden or costs to private companies.

Notwithstanding the vast quantity of scholarly work hitherto carried out on the evolution of environmental statehood in the EU, most researchers tend to describe the public sector as a consistent, predetermined entity put in charge, on behalf of the whole of society, of independently mediating social demands and environmental impacts. Because of the common assumption – associated with Kantian political theories – that the state is the representative of supreme rationality and an agent capable of understanding the demands of the entire society, mainstream interpretations of policies and policy-making have persistently struggled to explain the deeper dilemmas faced by the EU state system and the unexpected reactions of social and economic sectors. Many authors have repeatedly insisted on interpreting the state as a cohesive and predetermined entity that was put in charge, on behalf of the whole of society, of mediating environmental disputes between social groups and spatial units. The main consequence is that the prevailing readings of policy-making fall short of explaining the unsatisfactory results obtained with the introduction of policies and programmes supposed to reflect the best knowledge and regulatory skills of the state. For instance, the establishment of a network of around 25,000 conservation areas throughout the EU known as Natura 2000 sites is often described as one of the most important efforts in terms of environmental conservation so far. However, its legitimacy has been questioned by groups and communities involved in the designation of conservation areas, especially because its style of environmental protection invites a technocratic, top-down mode of policy-making, as demonstrated by serious problems in terms of communication and public engagement (Keulartz and Leistra, 2008).

Habermas (1991) had previously drawn attention to the contradictions of the European state system due to the requirement to become more inclusive at the same time as it loses its ability to operate effectively as a 'rational' entity. The instrumentalised rationality that underpins EU environmental statehood is clearly revealed through the use of the DPSIR framework for the formulation of public policies (the acronym stands for Drivers forces, Pressure, State, Impact and Response). DPSIR is widely adopted by OECD countries and, specifically, by the European Environment Agency to assess the causes and the evolution of environmental problems. Each term of the assessment framework informs the production of a range of indicators, which are supposed to facilitate communication to policy-makers and the general public (Gabrielsen and Bosch, 2003). Although the DPSIR framework has been increasingly employed in research projects to organise indicators in a meaningful way, there has been much criticism of its internal logic. Niemeijer and de Groot (2008) identify a simplistic, unidirectional chain of causality between the various sets of indicators and advocate the incorporation of multiple causal networks. Equally, Carr *et al.* (2007) see an excessive rigidity in dealing with drivers and responses at different levels. There is, however, a more fundamental weakness in approaches such as DPSIR, which is their profoundly anthropocentric, linear and ultimately authoritarian nature of assessment and decision-making. The exercise leaves room only for a highly controlled interpretation because it unfolds through a set of preordained steps.

The regulation of water use and conservation vividly illustrates the achievements and shortcomings of the flexible model of environmental statehood in the EU. The majority of the studies focused on water regulation in EU countries tend to concentrate on the superficial managerial adjustments and on the attainment of consensual responses. There is less interest in discussing the multiple contradictions and political clashes associated with the introduction of water legislation and the organisation of responsible regulatory agencies. The approval of the Water Framework Directive (henceforth, WFD) by the EU member states in 2000 was a particular milestone in the history of environmental regulation not just in Europe, but around the world. Because of the combination of environmental targets, economic safeguards and social sensitivity, the adoption of the WFD has been described as a major step forward in the contemporary search for better environmental regulation and water management. The WFD represents the latest stage in a sequence of international reforms that started years earlier and attempted to replace traditional approaches – largely based on rigid regulatory controls – with more flexible, adaptive and comprehensive responses to water management problems. Instead of single purpose, engineering-based initiatives, governments and society are now expected to systematically address freshwater extraction, effluent discharge and the physical alteration of water bodies (Ioris, 2012).

The implementation of the WFD has been also responsible for the growing politicisation of water regulation, associated with increasing controversy about the best way to accommodate conflicting interests regarding water allocation, use and conservation. Water policies and regulation have evolved through the prioritisation of some socioeconomic demands in a way that inscribes the balance of power in the management of water itself. Because of the significant costs involved in restoring the ecological condition of water bodies, there was a lengthy argument between the state, the market and civil society for the approval of the new Directive, which came in addition to prolonged bickering between the European Parliament and the Council of Ministers, intermingled with the pressures of different interest groups and NGOs (Kaika and Page, 2003). The main reason for the controversy is that water management issues are more than simply technical and physical questions, but encapsulate conflicting values and complex interactions between social groups and economic sectors. If the introduction of the WFD offered an opportunity to enhance water management and conservation, the implementation of the new Directive has been fraught with new challenges and shortcomings, such as the tendency towards non-compliance and the difficulty to adopt the requirements of WFD by some member states (Liefferink *et al.*, 2011), difficulties with defining regulatory standards and securing additional resources (Kanakoudis and Tsitsifli, 2010), as well as with the lack of commitment, leadership, public involvement and transparency (Watson *et al.*, 2009). WFD regulators have typically made use of only a narrow sub-set of options that replicate their previous (i.e. pre-WFD) approaches, and this is a result of resource constraints, lack of scientific data and institutional inertia, among other factors (Kirk *et al.*, 2007).

The identification of the state as the ultimate guarantor of rational, reliable responses to collective water problems has certainly shaped the WFD experience and a large part of the examination of results and failures. The shortcomings of most scholarly interpretations of the water institutional reforms are especially demonstrated by the failure to notice the biased mechanisms created for involving the public in the decision-making process. Public forums, such as the river basin groups, have been contained by sectoral interests and the rigid timetable of the WFD. It is normally the case that those groups that have historically controlled water allocation and use are still in control of the implementation of supposedly novel water institutions. State interventions are never politically neutral, but typically give rise to organised environmental destruction, systematic violence and sociopolitical domination. When confronted with the failures of the WFD regime, the reaction of politicians and academics has been to respond with more flexible and pro-active strategies described as the search for 'environmental governance'. Where the contested nature of the reforms is acknowledged, it is still from a very managerial perspective, as if water politics were a kind of deviation from the purist purpose of water management (McCulloch and Ioris, 2007). Most efforts are spent on designing fanciful computer models and assessment techniques instead of dealing with the politico-economic causes of environmental impacts or with the social and spatial asymmetries responsible for the unfair allocation and unsustainable use of water. The need to produce a regulatory framework that is both rigorous and malleable represents a formidable challenge for environmental regulation in the EU and demonstrates the persistent struggle to square the circle of private and public demands.

That has certainly been the case of the implementation of the Directive in Scotland, a semi-autonomous state that forms part of the United Kingdom and the European Union. The Scottish experience illustrates the limitations and intricacies of contemporary water reforms along techno-environmental, monetisation and legitimisation lines. Going a step beyond mere bureaucratisation, the reforms associated with the WFD have represented an invaluable opportunity to reaffirm the authority of the Scottish State, such as improving the effectiveness of other public policies on energy, agriculture, urbanisation and health. The transition from old to new regulatory approaches in Scotland has not been without tensions and inconsistencies. The implementation of the WFD has prompted a far-reaching renovation of policies, use and control of water, because it is the first time that water use is comprehensively covered by a single piece of legislation. Government agencies have praised the translation of the WFD into Scottish legislation in 2003 as a genuine opportunity to deal with water management problems in a country that depends socially, culturally and economically on the 'water environment'. The introduction of new environmental legislation involves the identification of (old and new) questions, the development of specific policies and the prioritisation of responses, which all necessarily require some form of political negotiation and intersectoral compromise. In the specific case of Scotland, this overall context of disputes and politicisation was further fuelled by the reinstallation ('Devolution') of a national parliament and executive government in 1999, which interestingly coincided with the late stages of preparation and final approval of the WFD.

After nearly three centuries of united history, since the Treaty of Union in 1707, Scotland regained control over a range of public matters, including overseeing the implementation of the WFD in one-third of the British territory (i.e. the area of Scotland in the UK). The movement for Scottish self-determination is not new and has evolved over the past decades in an interlocked process of identity definition and fierce struggle for economic recovery. In effect, Scotland had suffered more than other parts of the UK due to problems such as declining population, emigration, unemployment and extensive foreign ownership of local businesses, issues that politicians have repeatedly used as compelling arguments in favour of home rule (i.e. Devolution). In this context, questions related to water management are only some among many areas where Devolution still remains an incomplete process, fraught with overlaps and uncertainties. For instance, in the case of energy generation (such as hydropower, which accounts for 10 per cent of Scottish electricity), public policies on energy are still a prerogative of London, but the authorisation to build new schemes (under the planning permission regime) is granted in Edinburgh or by local authorities. The historical coincidence between the WFD and the reinvention of the Scottish Administration facilitated the convergence of water regulatory reforms with the broader reorganisation of public affairs.

It is well known that the reorganisation of environmental governance in Scotland was initiated a few years before Devolution with the approval of the Environment Act in 1995, which amalgamated various River Purification Boards under the Scottish Environment Protection Agency (SEPA). The creation of SEPA was already a response to the changing landscape of environmental regulation in Europe that came to require a more proactive role from public organisations. Administrative reforms in the water sector included the consolidation of the water industry into a single public utility (Scottish Water) in 2002, in which the same principles of administrative rationalisation and cost reduction used to justify the creation of SEPA were invoked. At the same time as the public sector was being transformed, Scottish representatives were heavily involved in the negotiation of the new Water Directive, in particular, because the last chairman of SEPA (Sir Ken Collins; in office between 1999 and 2007), then a Scottish Member of the European Parliament, was in charge of the Environment Committee of the European Parliament and strongly supported the consolidation of European legislation into a single, comprehensive directive (Jordan, 2000).

The transition from a previously centralised UK Government to a 'devolved' Scottish administration has indeed had important material and symbolic consequences for dealing with water problems in Scotland. Before Devolution, it was significantly more difficult to reform Scottish law due to a shortage of parliamentarians' time (in Westminster) and the restricted relevance of Scottish issues in the UK political arena. This changed after 1999, when the new Scottish Parliament was able to mobilise time and resources for a comprehensive review of the existing water legislation. More than merely a historical coincidence, the fact that Scotland was the first region in Europe to translate the WFD into national legislation (sanctioned under the Water Environment and Water Services Act in 2003, ahead of the official deadline) reveals the political importance given to the water institutional reforms. In a short period of time (between 2005 and 2006), over 7,000 water use authorisations were issued by SEPA, ranging from simple to complex registrations, and multi-site licences. As claimed by the Scottish Government (2008: 06), "Scotland is at the forefront of influencing European policy on implementing the WFD, playing an important role in a range of working groups established by the European Commission".

The new parliament not only managed to produce a thorough legal reform that in some aspects goes beyond the requirements of the WFD (such as conservation objectives of coastal waters up to three nautical miles and the introduction of specific requirements to identify and monitor pressures and impacts in wetlands), but also benefited institutionally from the political significance of having to translate the WFD into national law. In other words, the priority given to the WFD by the Scottish Parliament was not only a chance to improve water legislation, but also contributed to the very affirmation of the newly 'devolved' parliament. In fact, the early approval of the WFD was praised as a demonstration that Scotland can do things faster than the rest of the UK and that the implementation of the WFD in Scotland has been both timely and systematic.

In articulating a sense of national pride around the forthcoming water legislation, the young Scottish administration systematically attempted to assert its authority by forging a range of channels with the main water user sectors. Nonetheless, while the involvement of some groups of stakeholders played an important role in shaping the new legislation, it did not necessarily result in stronger democratic representation or better environmental governance. On the contrary, lobbying and bargaining around the adoption of the WFD have exposed a highly controlled process of public involvement and stakeholder contribution. Ison and Watson (2007) made it clear that the approval of the new law was basically the product of a handful of officers from the Scottish Government and parliamentarian advisers, who worked closely with three representatives of NGOs (known at that time as the 'three witches'), with wider consultations coming only later in the process and, crucially, when most decisions had already been made. The selective basis of participation continued throughout the implementation of the WFD in Scotland, especially because the main institutional mechanism for involving the public has been public consultation. The obvious weakness of asking the public via formal consultations is that official agencies have ample discretion to accept or reject any suggestion received through the process.

As mentioned above when describing the WFD experience in general, since the early years of the new Scottish legislation, there has been a persistent difficulty to genuinely consider the inputs of stakeholders and local demands. For example, the design of the River Basin Districts, which are the administrative units of the new regulation, involved a series of meetings in different parts of Scotland in 2003. Nonetheless, while the public seemed to favour a format of River Basin Districts that coincided with catchment boundaries, decision-makers preferred to group unrelated catchments under the same district area. At that point most of the public also insisted, to no avail, on a more flexible and realistic timetable to implement the new Directive, which would have allowed more time for raising awareness and debating water problems. With extra time available, a number of local initiatives, such as catchment and stakeholder mobilisation schemes that existed throughout Scotland, could have better informed the preparation of the coming regulation. In addition to public consultations and some ad hoc seminars, which happened especially between 2002 and 2005, the other opportunities for involving the public in the debate have been related to the preparation of the River Basin Management Plans (the strategic decision-making process introduced by the WFD). The Plans were discussed regionally by ten Area Advisory Groups (AAG), which are the official forums of public debate and sectoral negotiation. Despite their democratic appearance, the activities of the AAGs included a series of meetings with a rigid timetable and little flexibility for unexpected, time-consuming controversies. AAG representativeness is further weakened by the fact that its membership is decided unilaterally by SEPA, and the role of its members has traditionally been informative rather than operational. In other words, the scope of AAGs in Scotland was basically restricted to fine-tuning the production of the River Basin Management Plans (RBMP), instead of really engaging with the decision-making process. There were increasing complaints that the RBMP experience in Scotland is, by and large, a 'tick the box' exercise of the official agenda of the implementation of the WFD, together with frustration at the lack of willingness on the part of SEPA to share data and discuss internal technical procedures. It is not entirely clear how the agency assessed the individual and cumulative impact on the environment caused by different water users in the same area. Consequently, there were limited prospects for members of the AAG to influence how SEPA would deal with the mitigation of water management problems.

At the same time that a significant effort was being made to try to conform to the European legal requisites, water management problems continued to emerge in various parts of Scotland. For instance, in the Loch Tay area, in the southern Highlands, there was growing competition between hydropower developers, local water supply operation and environmental conservation objectives. This conflict has been recurrent in various applications to build new hydropower schemes in the Perthshire area, where the uncertainties related to the implementation of the new regulatory regime, together with the inconsistencies between planning development and water regulation, have created uneasiness among local and national stakeholder groups. Given the current policy of the Scottish Government in favour of small and medium-size hydropower schemes, this kind of dispute is likely to increase. Similar contentions exist between sites affected by the construction of new dams (especially for urban water supply and hydropower) and the remote areas expected to benefit from additional electricity and public water supply. It is worth mentioning the existence of an intricate network of pipelines in the Highlands of Scotland that serve to transfer water from one catchment to another. Decisions on the management of these multi-site schemes only add a new layer of complexity to an already complicated regulatory regime. Likewise, in the catchments shared between Scotland and England, water management has become entangled in a not-always-easy relationship between public agencies to the north and south of the English border. The sub-national experience of the WFD in Scotland can be compared here with other devolved administrations in Europe, such as in the Lower Saxony, where the implementation of the new Directive has depended on a series of contextual and contingent inter-regional issues within the nation-state (Kastens and Newig, 2007).

The influence of the stronger players on the implementation of the WFD in Scotland has thwarted the more innovative prospects of the new regulatory regime, such as the modification of the overall pattern of wasteful water use by households and business sectors. The largest water users – above all, hydropower, farmers and the public water company – have constantly exerted their political leverage to shape public policy in order to maintain business interests and ensure that everything remains unchanged. The stronger sectors have even managed to secure an exclusive agenda of discussions with the regulatory agency, which has not always been sufficiently transparent to the other parties concerned. The liability for environmental damages, such as in the case of the whisky distilleries that diverted entire streams to serve their water needs, was systematically denied with the claim that economic results are more important than trying to restore the river systems. Although the WFD, similarlsy to political Devolution, was promoted as heralding 'new politics' of democratic recovery via a more open approach to public matters, the actual practices of governing the environment continue to owe more to traditional rationalities of centralised managerialism (Thompson, 2006). A clear manifestation of such rationalities has been the heightened role played by environmental economics in the implementation of the WFD in Scotland.

A key policy instrument of the WFD regime is the requirement of all water users to make a payment equivalent to the environmental impact they create (normally described as the 'polluter-pays' or 'user-pays' principle). The compensation for the negative impacts on the environment takes the form of bulk water charges to be paid regularly to the regulatory agency (SEPA, in the case of Scotland). According to the literature on environmental economics that underpins most of the new institutional arrangement, bulk charges serve to internalise the social costs (i.e. negative externalities) of using the environment for private benefits, thereby introducing an economic rationality that stimulates the efficient and sustainable use of water. In Scotland, the income from the new charges was expected to cover 50 per cent of the operational costs incurred by SEPA, while the other 50 per cent was to come from the government in the form of general taxation (SEPA, 2005). The introduction of the WFD charges was the object of two specific public consultations carried out by the Scottish Government in the year 2005. The first dealt with charges to be applied during the initial phase of issuing the WFD licences (described as the 'transition period' of 2005/2006) and only received seventeen responses (from green NGOs, such as the World Wide Fund for Nature and the Royal Society for the Protection of Birds). The second consultation in 2005 dealt with the full-charging scheme (i.e. the one supposed to be in place after the transition period) and attracted 189 responses. The main user sectors were clearly against and fiercely expressed their apprehension about the new charging mechanism. This second consultation process took place at the same time as the beginning of applications for the new WFD authorisations, which only added animosity to an already contested process.

To understand the meaning of bulk water charges in the Scottish experience (which is similar to the controversies in Brazil discussed in the previous chapter), it is important to consider that instead of facilitating the implementation of water reforms the principle of cost recovery has entangled SEPA in a hostile environment of lobbying and public disapproval that corresponded to the most turbulent period of the WFD regime to date. SEPA was seen, particularly in the mass media, as a draconian agency that was trying to sell the new regulation to secure its financial survival. This opinion was frequently repeated by individual stakeholders, especially those who were required to apply for a WFD authorisation to maintain current uses of water. Under serious criticism, SEPA had to quickly respond via a number of unscheduled meetings and ad hoc negotiations with water user sectors. The hurried amendments to the abstraction charges, during and after the consultation, demonstrate the concern that abstraction charges have caused among farmers and large users. SEPA actually had to make several concessions during the development of the charging scheme, which will probably come back to haunt the agency in the future. For example, the agency was forced to include several technical indices in the calculation of charges, which can now be challenged by the user sectors; likewise, the 'elastic' definition of abstraction points, which for irrigators can extend for many miles and can overlap with other water users, will be a likely source of conflict between farmers (e.g. the majority of potato growers rotate their equipment across different catchments and are likely to compete for the same stocks of water in the years when the areas of production coincide in the same catchment).

Because of the controversy surrounding bulk water charges (which was probably inevitable, given that hitherto, water use was often a free right attached to land ownership), the initial implementation of the WFD in Scotland was primarily associated with the economic dimension of water, at the expense of other initiatives more directly related to the mitigation of environmental and social problems. Although the payment of charges for water use is obviously part of the new regulatory regime, to a large extent it has become an objective in itself, particularly because the activities of the environmental regulator depend on the successful collection of those charges. In conceptual terms, it means that environmental objectives attached to payments became increasingly subordinated to the financial sustainability of SEPA. This is even more serious considering that the entirety of the water charges is used to finance half of the regulatory costs of the agency to oversee the implementation of the WFD. Unlike the water charges of other countries, such as France, where the income is reinvested into the improvement of the water environment, the charging scheme in Scotland represents a significant deviation from the objectives of applying the polluter-pays principle. Not only were new charges not directly related to investments in environmental restoration, but they also had the negative consequence of reducing the multiplicity of social and natural values of water to the single dimension of money value.

Among the various user sectors, the public water industry was probably the one that reflected this gradual commoditisation of water most clearly. As mentioned above, in 2002, the consolidation in a single utility company (Scottish Water) was presented as the best operational alternative to avoid the persistent financial losses of the then three water companies. Continuous losses had then placed the market value of the Scottish water industry at, at least, £500 million less than its outstanding debt (WIC, 2007). Since the reorganisation, Scottish Water has recovered its regulatory capital value, after having invested £413 million and achieved cost savings of more than £1 billion (Scottish Water, 2007). Not only is its management increasingly driven by an economic rationale, but the public image of Scottish Water has also been dominated by pressures to reduce costs and improve performance. Additional legislation, such as the Water Services Scotland Act 2005, required Scottish Water to establish a separate retail entity to compete on a level playing field with new entrants in the water services market. Under mounting pressure, the company moved beyond its solely governmental status and created a new commercial branch – Scottish Water Solutions – a joint venture between Scottish Water (51 per cent of shares) and two consortiums of engineering and private water companies (24.5 per cent of shares each). Scottish Water Solutions was required to deliver 2,500 projects estimated at £2.3 billion with a budget of only £1.81 billion (in other words, deliver more with less money). However, these financial savings were not immune to criticism, especially considering that the selection of investment priorities is largely subject to political and commercial pressures. In effect, interventions have been concentrated in locations and catchments where there are higher profits for the companies that form Scottish Water Solutions or where the achievement of targets is relatively easier. The use of 'rational' analytical tools to select and justify water management decisions is certainly not new, but the emphasis on this kind of approach has encouraged the subordination of the institutional water reforms to commercial targets and business strategies.

Another aspect of the controversy around the economic dimension of water is related to disputes about the costs of mitigating environmental impacts. The WFD is, by definition, a 'framework' type of legislation, which means that it systematises the direction that European countries should follow, while details of the application are delegated to the national administrations. Within reasonable technical boundaries, member countries can interpret the requirements of the Directive in order to restore water bodies to 'good ecological status'. If the current condition deviates from a good status, a series of measures must be put in place to guarantee environmental restoration by 2015. To inform the achievement of 'good ecological status' in Scotland, a series of publications have tried to calculate the monetary value of environmental conservation and the cost of restoration. Under the WFD the regulator can only impose mitigation measures that are 'feasible' and 'proportionate'. The financial costs associated to environmental compliance can be avoided, or at least minimised, if the activity is considered eligible for 'derogation' (cf. Article 4 of the WFD; see European Commission, 2000). Activities that cause serious environmental impacts can lawfully continue to operate on grounds of disproportionate costs, public interest or sustainable development goals (i.e. the criteria for derogation). Although the mechanism of derogation, if not well used, can undermine the rigour of the new regulatory regime, it was sometimes manipulated by SEPA as an appeasement strategy to remove opposition to the translation of the WFD into Scottish legislation. In the early days of the implementation of the WFD, some SEPA representatives even tried to persuade antagonistic voices to accept the new regulation under the argument that the really decisive phase would only come later, such as the assessment of environmental risks and the appraisal of derogations. In practice, because of the difficulty to please all social groups, the promise of a comprehensive assessment and detailed mitigation measures was increasingly frustrated due to political pressures. SEPA was led to disregard some onerous mitigation options and had to concentrate on a relatively small number of 'cost-effective' (i.e. normally less stringent) alternatives.

Arguments about the reasonableness of restoration costs continued to poison the dialogue between water stakeholders and the environmental regulator. In such situations of conflict, economic assessments are often used to protect established interests, such as the expansion of hydropower generation (incidentally, Moran *et al.*, 2007, projected an increase in freshwater use by 39 per cent between 2001 and 2015, primarily driven by hydropower generation). A concrete example was the dispute about the 13 miles of the River Garry (in the centre of Scotland) that lies dry most of the year because water is held back by a hydroelectricity dam. Environmental economists compared the cost of power generation with the marginal benefit of increased revenues from fisheries and concluded that the 'social' costs of reducing electricity are disproportionately high in relation to the value of fish preservation. The local fisheries board obviously disagrees and continues to claim that the River Garry is "Scotland's most abused river" and that the "electricity produced by the Tummel–Garry (hydropower) scheme was intended for sale to the towns and cities in lowland Scotland as opposed to Highland communities".

The Scottish experience also comprised a number of cases where the interaction between private stakeholders and the public sector was more balanced and transparent than usual (such as the support to the development of the catchment plans of the Dee and Tweed rivers, and the attempt to discuss the new regulations with irrigators in some small catchments along the east coast). These isolated initiatives were, however, exceptions. On the whole, the opportunity to promote a new paradigm of water regulation, beyond the priorities and the constraints of the neoliberalisation of socionature, was largely missed in Scotland. Instead of implementing the WFD through innovative forms of dialogue and cooperation, as claimed in numerous policy documents and official speeches, the new water regulation has followed the wider model of nature neoliberalisation that prevails in the European Union. Since its early days, different authors indicated the existence of serious obstacles for the success of the WFD in Scotland (e.g. lack of regulatory authority and enforcement resources); and only a few years later, it was already possible to identify a broadening gap between the ambitious rhetoric and the narrow reality of the new water regulation. In effect, several stakeholders contacted during this research (particularly those involved in the preparation of the River Basin Management Plans) have voiced their increasing frustration with the repetition of old practices and mistakes of previous environmental regulations.

This complex institutional and political landscape reveals a great deal about the new model of environmental statehood under implementation in Scotland and in the European Union at large. At face value, the range of events, reports, consultations and media coverage concerning implementation of the WFD in Scotland may give the impression of an ample reform. It seems that long-term management problems, such as degraded river stretches, obstruction of rivers by large dams, low river flow during dry summers and declining fish population will be finally resolved; or that the government is employing economic incentives, including fees and taxes, to manage water systems according to broader public priorities that are widely discussed with the population. However, a more careful analysis betrays the superficial level of change and the overambitious rhetoric of the WFD. It becomes clear that the actual methods of assessing problems and formulating solutions have mostly reproduced the patchy responses that existed before the WFD. What is more, the timid scope of the ongoing water reforms in Scotland has eroded the prospects of effectively improving the use and conservation of water and reversing anti-commons tendencies. For instance, despite all the effort involved in the new charging scheme, it basically aims to recover the operational costs of SEPA rather than to internalise the social costs of water use. To many who took part in the public consultations and protested in the media, the WFD regime has been focused on imposing fees and taxes rather than on environmental conservation.

The growing frustration of water stakeholders with the WFD in Scotland certainly echoes the contested nature of the international experience with the adoption of flexible environmental statehood. What is peculiar to the Scottish experience, though, is the fact that, because of the reconfiguration of Scottish public affairs, public involvement and economic-based regulation have evolved during the unique circumstances created by Devolution. The reform of water institutions in Scotland provided a chance to understand the connections between territorial politics, environmental vulnerability and economic pressures. The WFD is a legal requirement that was opportunistically transformed into a compelling argument in favour of the benefits of the self-government of Scotland. The political construction of Devolution has triggered an asymmetric territorialisation of water policies within Scotland and between different parts of the United Kingdom, and has also exacerbated the economic dimension of water use. While most user sectors (agriculture is the exception) are likely to increase their use of water significantly during the implementation of the WFD, the new regulation has been incapable of dealing with the close relationship between poor water quality and social deprivation in marginalised areas, such as the suburbs of Glasgow and Edinburgh.

The reliance on the generic assessment of ecological processes and the quick-fix solution to long-term environmental impacts betrays the irrational rationalism of flexible environmental statehood. Because of its technocratic heritage, the key outcomes of the WFD experience have been an artificial (and mostly unnecessary) complexification of water management and the widespread use of the money language: the dominant forms of dealing with water remain bounded by economic assumptions about how nature operates and how natural resources should be used. The 'cash nexus' inevitably results in the hypertrophy of the economic dimension of managed water systems, at the expense of other social and cultural characteristics, at the same time as there is no empirical evidence that monetisation improves environmental management. On the whole, the failure to articulate alternatives has left water reforms in Scotland exposed to the neoliberalisation of public policies that ultimately constitutes the fabric of contemporary approaches to environmental problems. Because of the powerful influence of neoliberalism, the new regulatory regime may be capable of acknowledging complexity and sectoral demands, but it is reticent when it comes to allocating responsibilities for environmental impacts. The standard assessment of environmental impacts effectively dilutes the responsibility for the genuinely serious damages, inasmuch as it basically deals with the most common impacts and only resorts to a limited range of mitigation responses. Such a conclusion should not come as a surprise, since it is a characteristic typical of neoclassical economics, strongly related to the WFD experience, to say little about the evaluation of past mistakes in relation to the environment.

Overall, the two strongest institutions advanced by the WFD, namely 'the search for economic efficiency' and 'the selective involvement of the public', have not produced fundamental changes in the forms of using and conserving water in Scotland. Although some localised and patchy improvements are expected as a result of the WFD, the introduction of economic-based regulation will continue to raise conflicts and contradictions (as in the case about deciding derogations). This is due to both the WFD being entangled in multi-levelled disputes and, more importantly, the WFD being in itself a very limited piece of regulation that, despite its formal requirements, has perpetuated an uneven and centralised management of water use and conservation. The shortcomings of the water regulation reform in Scotland indicate a pressing need for deeper changes in environmental statehood, which should be strongly grounded in principles of environmental justice and positioned beyond standardised solutions to water management problems. Because of the intrinsic politicisation of water management, effective answers to old and new problems continue to require more inclusive decision-making and a schedule of discussions and negotiation designed to be more realistic than the rigid and predetermined timetable of the WFD. By the same token, public policies should avoid positivistic or technocratic approaches to water management problems and recognise the full extent of the complex relationship between society, state and the rest of socionature.

The contingent relation between the search for flexible environmental statehood and the subtle process of nature neoliberalisation in Scotland has important parallels with the reorganisation of the Peruvian State and the neoliberalisation of public water services in Lima, as discussed next.

Urban socionature and environmental statehood changes in Lima, Peru

The complex transition to a more flexible configuration of environmental statehood through the growing neoliberalisation of nature was not exclusive to European Union countries; the same process of institutional change and intensification of anti-commons trends took place in many other parts of the world. A comparable example is provided by the introduction of new approaches for the management of public services and nature conservation in large metropolitan areas. In view of their mounting importance, urban questions are right at the centre of the contemporary debate about state, society and the rest of nature. The 'urban' has become a main sphere of social activity, political contestation and capital accumulation, which all happen through relations that are profoundly socioecological. Megacities encapsulate the key challenges for the state in terms of social cohesion, environmental conservation and the globalised economy. For instance, the growing number of Latin American metropolises is characterised by pockets of urban wealth and ostensible affluence amid vast areas of deprivation, overcrowding, pollution and multiple forms of violence. In the last half century, medium- and large-scale cities in Latin America underwent an accelerated process of expansion fuelled by internal migration and high rates of population growth. The consequence is that Latin America is today the southern region with the largest proportion of the population residing in metropolises of over one million inhabitants (32 per cent of the Latin American population, compared with 15 per cent for Asia and 13 per cent for Africa, according to Cohen, 2004). This uneven pattern of urbanisation is, ultimately, the result of several decades of national development subordinated to the narrow interests of the middle classes and the small governing elites.

The evolution of large metropolitan areas, particularly in Latin America, can be explained in relation to several interlinked processes that permeate public and private realms. The consolidation of large conurbations represents only the most recent chapter in the long trajectory of nation building and economic production which has been orchestrated by the state apparatus. In addition, because of the elitist and subordinate pattern of national development, the large cities have become the main arena for the crystallisation of socioecological inequalities and the persistent reinforcement of environmental injustices. National development has relied on the large-scale cities to accommodate industries, the labour force, the techno-bureaucratic administration and a myriad of technological and financial services. In the process, significant contingents of the poor population have been attracted to the megacity, but only a small proportion of them have been absorbed into the formal urban economy (i.e. mainly through the sale of their labour power). On the other hand, the mega Latin American cities are today the main locus of mobilisation, creativity and political action, as well as for the experimentation of both top-down and bottom-up responses to collective problems. The 'negation' by the megacity, through the persistence of structural inequalities, is then negated by those that live in the political and spatial periphery, who need to resort to alternative strategies to cope with deficient public services, widespread violence and institutionalised forms of mistreatment. The creativity and energy of the urban poor do not eliminate their exploitation, but are a fundamental element of their fight for political rights and better living conditions. The negation of the negation becomes a new affirmation, or at least a utopic possibility, of overcoming the failures produced through established forms of urban expansion.

The circumstances of Lima, the capital of Peru, demonstrate the complex dilemmas faced by the national state in relation to mega urbanisation. The problems of metropolitan Lima have their origin in the long history of exploitation and negligence perpetrated against nature and society since colonial times. The founding of Lima in 1535 was not only a milestone in the organisation of the new colony, but it became the main asset for the expression of Spanish power over the local inhabitants and the rest of socionature. In contrast to the autochthonous Peruvian nations, who developed most of their settlements in the mountains where water reserves were more abundant, Lima was established in an area with very limited rainfall and only three small watercourses (Rímac, Chillón and Lurín). The main reason was the need to exercise power from the coast in order to facilitate communication with Europe and protect the colonial authorities from the threats of the dominated peoples. For centuries, Lima remained the main urban centre of the Spanish territories in South America and had a critical role in the control of both transatlantic and transpacific trade routes.

Around the turn of the twentieth century, Lima was a complex mosaic of rich and poor households intermingled along the same streets, with large mansions, middle-class houses and crowded working-class residences lying side by side. Gradually, however, the wealthier groups started to leave the oldest parts of the city and migrate to the southern areas along the coast. During most of the century, there was an increasing concentration of the Peruvian population in Lima due to the influx of people from other coastal areas and, later, from the Andean mountains. Lima operated as a magnet that attracted large contingents of people from the provinces. Poverty and lack of space provided strong incentives for the creation of alternative forms of accommodation (especially in precarious urban settlements called *barriadas*). With the mounting scarcity of land, the new barriadas had to be established in distant and hilly areas where public services and civil rights were even less attainable.

Those tensions have deepened in the last two decades, when the socio-spatial configuration of Lima was profoundly influenced by neoliberal strategies. At the same time, the everyday life in the large city was shaped by manifold reactions to the lack of spatial and social opportunities. Lima was set on a trajectory of mass consumption and conservative ecological modernisation, at the same time as social inequality and underemployment proliferated. The capital encapsulated and magnified the expanding neoliberalisation of Peruvian society, a process that started in the 1980s with macroeconomic changes, intensified in the 1990s with the privatisation of most public utilities and continued in the following decade with the private appropriation of cooperatives and common land in remote parts of the country. The neoliberal megacity is increasingly hierarchical and exclusionary and, consequently, it shows much continuity with the previous phases of urban development. Urban integration through mass consumption, unstable jobs and utility privatisation has been unable to respond to the needs of the majority of the urban population, but serves primarily the aspirations of middle and upper classes. Likewise, the weakening of community mobilisation is an adjunct of the expansion of neoliberal urbanism and helps to reinforce urban trends characterised by acute individualism. Notwithstanding investments in the modernisation of affluent areas and large infrastructure projects, for those living in the barriadas and in peripheral neighbourhoods the access to public services and a reasonable quality of life still remains a daily battle.

The introduction of neoliberalism and its significant impacts on statehood followed the repercussions of the nationalist and highly incompetent government of Alan García in the 1980s. Ill-conceived, populist interventions led the country into a period of hyperinflation, generalised instability and economic depression. That dramatic process of change is vividly described in the *mémoires* of the novelist and former presidential candidate, Mario Vargas Llosa (2005), defeated in 1990 by Alberto Fujimori. After his unexpected election, Fujimori embraced Vargas Llosa's neoliberal manifesto of economic recovery through market deregulation, state retrenchment, normalisation of debt service payments and reinsertion of Peru into the international financial community. Fujimori came to office with practically no coherent plan of action and was led to adopt a draconian programme of state reforms, privatisation and economic adjustments. Fujimori dismantled the existing mutual system (under the administration of the Housing Bank (*Banco de la Vivienda*), which had mostly supported investments in the middle-class areas of the city) and other assistance mechanisms for the low-income neighbourhoods and the barriadas. The key decisions about urban development were progressively centralised in the hands of the national government at the exclusion of the forty-nine municipal authorities that govern the metropolitan area of Lima. With the 1992 self-coup (*auto-golpe*), the regime became semi-dictatorial – including control over the congress and the judiciary – which was instrumental for the wider neoliberalisation of the economy. The eventual resort to violence and repression by the Fujimori administration seemed to confirm the early observation of Slater (1989) that democracy may not be the best political environment for capitalism in peripheral countries such as Peru.

A new national agency (called COFOPRI) was established in 1996 (with financial backing from the World Bank) with responsibility for granting land titles and regularising informal settlements. Fujimori systematically manipulated the granting of titles by COFOPRI according to his electoral needs, especially because it was easier to secure votes in the crowded periphery of the capital than in the remote provinces. COFOPRI's purpose was directly informed by the ideas of De Soto (1986), who had claimed that the formalisation of land tenure would energise commercial transactions in the city. Following this ideological position, COFOPRI became a concerted attempt to stimulate the circulation of capital through the concession of loans for the acquisition of family properties. In practice COFOPRI also created an opportunity for real estate barons and commercial banks to siphon off public funds, especially because of home loan foreclosures and the displacement of families. Between 1996 and 2000, around half a million property titles were granted in Lima, but at the same time the number of protests aimed to establish new barriadas was never so high. The total number of barriadas had reached 1,980 in the year 1998 and included 2.6 million inhabitants or 38 per cent of the population of Lima (Calderón Cockburn, 2005). By contrast, in the central areas of the city, the neoliberal renovation resembled the experience at the turn of the twentieth century, when Lima was remodelled under French and English aesthetic influences to satisfy the demands of the wealthier strata of the population. The most emblematic construction of the neoliberal phase was the shopping centre Larcomar, built in the scarp of Miraflores in 1998 with a total investment of more than US$40 million. High-income residences and service offices were increasingly accommodated in multi-storey buildings (e.g. international banks and companies in the San Isidro neighbourhood), while the low-income areas of the city continued to expand horizontally and up the hills.

For many decades, from the interwar period to the neoliberal phase, successive governments tried to deal with the fast rate of metropolitan expansion and the spread of the barriadas. However, the effort largely failed to stem the chaotic growth of the city to the north, the south and the east, especially along the main roads and river valleys. Successive initiatives also did little to prevent the consolidation of an overarching pattern of deprivation and uneven development between central and peripheral neighbourhoods. An important aspect of the pressing urban problems of Lima in the twentieth century was the slowing down of investments in public water services, which was aggravated by escalating water demand, inadequate planning and the deterioration of the infrastructure. Because of many decades of mismanagement of the sector, after taking office as president, Fujimori inherited a city on the brink of a water crisis. That grim situation was ultimately the legacy of the García administration in the 1980s which had compromised the managerial and planning ability of government agencies. The water utility of Lima (SEDAPAL) had been plagued by unceasing political interference and was frequently criticised for its distant, unresponsive relationship with the population. More importantly, the need to reorganise the public water services coincided – and had a contingent relationship – with the introduction of pro-market institutions and the creation of new avenues for capital accumulation under Fujimori.

From 1990 the water industry of Lima became an important arena for testing the technical, economic and political dimensions of water neoliberalisation mentioned earlier in this chapter. The experience of Lima not only provides compelling proof of the political willingness of the national elite to reform the Peruvian State according to liberalising goals, but also demonstrates how multiple socio-ecological reactions forced systematic adjustments in policies and procedures. The introduction of water neoliberalisation in Lima came slightly later than in other neighbouring countries (such as the early initiatives in the State of São Paulo, in Brazil in the 1980s) and was only possible after the general election won by Fujimori. Despite this delay of a few years, the initial focus of the reforms was on the reorganisation of SEDAPAL with the purpose of achieving higher rates of technical and managerial efficiency, which corresponds to the techno-environmental dimension of the analytical framework. Because of the critical condition of public water services, the incoming Fujimori administration was initially forced to implement an emergency plan for water supply and infrastructure rehabilitation, which involved the construction of a number of small boreholes and storage tanks in low-income zones. Specific measures were taken to secure leakage reduction and some localised decontamination of the River Rímac. Transnational actors also played a very important role in the transference of know-how and institutional strengthening (e.g. the German programme Proagua since 1996 and the World Bank's Water and Sanitation Programme since 1995). In terms of state-fix and service regulation, SUNASS was established in 1992 as a dedicated agency responsible for overseeing the separation between policy-making and utility management, as well as for operation benchmarking and the enforcement of more stringent water pricing mechanisms.

After those technical efforts to improve service performance and contain environmental degradation, the next main step was to incorporate SEDAPAL into the agenda of utility privatisation then eagerly promoted by the Fujimori government. Alcázar *et al.* (2000) emphasise that privatisation (in this case, a concession to private operators) first required a careful reshuffle of the water utility of Lima, such as the review of water tariffs (increased from US$0.17/m^3 in 1990 to US$0.41/m^3 in 1995), the reduction of labour costs (between 1991 and 1992 the company lost 721 workers or 23 per cent of the workforce) and structural investments (an increase in annual investments from US$26 million in 1990 to US$80 million in 1996). Those adjustments benefited from a World Bank loan of US$600 million that specifically aimed to guarantee the commercial viability and public image of the water utility in anticipation of the intended privatisation. In a decade when the management of water utilities was being rapidly transferred to the private sector, the announced privatisation of SEDAPAL certainly attracted worldwide interest. In 1994, three consortiums formally expressed the intention to bid to take over the operation of the water services of Lima, namely Canal de Isabel II, Compagnie Generale des Eaux and Lyonnaise des Eaux. However, despite the apparent favourable policy environment, the desired privatisation of the water utility of Lima never happened. Following various delays, the tender was postponed until after the re-election of Fujimori in 1995, followed by further adjournments and, eventually, an official cancellation in 1997.

The main problem then faced by the Fujimori government was political, as the legitimisation of utility privatisation was then insufficient to overcome growing opposition by water users and civil society representatives. Ugarteche (1999) points out that the tensions related to the 1990s neoliberal reforms were effectively an attack on the rights and achievements of the working class, which inevitably raised opposition and, whenever possible (considering the authoritarianism of Fujimori), were resisted by the population. The public was particularly dissatisfied with the fact that privatisation would be followed by the significantly higher tariffs needed to recover the investments required from the private operators. With the momentary impossibility of privatising SEDAPAL, the government undertook a large programme of operational rationalisation and economies of scale. It is perhaps ironic that the administration of Fujimori, probably the most neoliberal government on the continent at that time, was directly in charge of a comprehensive package of investments in equipment, technology and construction contracts estimated at around US$2.44 billion in Lima alone (which was the equivalent of 0.5 per cent of the GDP of the entire 1990 decade; cf. SEDAPAL, 2005). Better management ended up alleviating the water problems and further reduced the appetite for privatisation within the national government. A set of mechanisms set in motion in an attempt to neoliberalise nature can sometimes "lead to events that, in turn, may modify or hinder the policies that brought about the initial change" (Castree, 2008b: 162).

For those living in the barriadas and low-income neighbourhoods, access to public services and a reasonable quality of life were still major problems at the end of the Fujimori administration. In 2001, only 11 per cent of the settlements regularised by COFOPRI had acceptable standards of public services (considering water, sanitation, streets and construction material for houses), according to SASE (2002). Because of escalating levels of crime, city enclaves in the form of gated communities have become a common feature both in high- and low-income areas of Lima. One of the significant results of the neoliberalisation of the economy was the deterioration of the levels of income of the workforce in Lima between 1987 and 2002, especially among non-unionised, independent workers (Verdera, 2007). Economic stabilisation happened mainly through the reduction of state expenses and extensive utility privatisation, although it also created a persistent mismatch between economic results and sociopolitical demands. The last years of Fujimori's administration were famously marked by massive corruption and mafia-like operations carried out personally by the president and members of his cabinet. Amidst growing scandals, the government crumbled in 2000 and was followed by the caretaking administration of Valentín Paniagua. However, the reestablishment of civil liberties and formal democracy by the new regime was not followed by changes in the underlying direction of the economy.

On the contrary, the neoliberal tide was resumed by President Toledo (2001–6), whose government was marred by constant political turbulence due to a parliamentarian minority and the adverse global economic situation (Murakami, 2008). Toledo came to office with the promise of overcoming the shortcomings of the previous governments that had left the capital fraught with institutional uncertainties, poor policy coordination and growing environmental impacts. The new state fund *MiVivienda* started to finance the purchase, improvement and construction of popular housing. Other projects and plans were also launched with the purpose of alleviating the housing deficit (e.g. *Techo Propio, Bono Familiar Habitacional*, etc.). However, the perverse consequence of those initiatives was the over-reliance on the private sector for the construction of new housing units, while the state largely withdrew from direct construction interventions. Under free market competition, the builders obviously showed a preference for middle-class residences instead of the less profitable units for the low-income population. The fact that it was increasingly difficult to identify a physical and symbolic centre for Lima can be interpreted as a metaphor of the barriers to promoting coordinated urban policies. Lima has various isolated centres (e.g. Cercado, La Molina, Miraflores, San Isidro, Callao/airport), which demonstrate the fluid configuration of power and money determining the functioning of the megacity.

In 2006, in what is one of the most curious turns of contemporary Peruvian politics, Alan García, the same leader who undertook a histrionic confrontation with the international financial system in the 1980s and even attempted to nationalise the banking sector, was returned to office as a converted neoliberal politician. The odd dialectics of García(1)-Fujimori-García(2) – in the sense that the second mandate of Alan García incorporated the neoliberal platform of Fujimori and blended it with his distinctive populist attitudes – only make sense in the context of patrimonialism, economic instability and weak political parties that have characterised the recent history of Peruvian development. During the campaign and throughout his mandate, García sustained his promise to remain faithful to the neoliberal canon. The economic reasoning of the new administration was bluntly revealed in a series of newspaper articles, published in October and November 2007, when the president blamed those against the neoliberal reforms for suffering from *el síndrome del perro del hortelano* (translated as 'the dog in the manger syndrome'). García criticised the fact that large extensions of land were being used by the peasantry, what was perceived as a lost opportunity for economic growth. Instead of leaving land and resources in the hands of peasants, García called for an intensified exploitation of water, gas and timber by national and international corporations. The administration took numerous measures to put the ideological claims made by the (recently converted) President García into practice. Furthermore, in December 2007, after the ratification of the Free Trade Agreement with the USA, the congress delegated to the executive – something that has not been uncommon in Peru – the express authority to legislate for six months over matters related to that agreement. That was García's 'Eighteenth Brumaire', with 102 decrees issued unilaterally by the president, including Decree 1,081 that replaced the previous water law with further legal reassurances for the operation of private sector investors.

The result was that the use of urban planning to assist the interests of private investors followed an even more distinctive trend during the second term in office of President García (2006–11). García reinforced the pro-market strategies of Toledo, which in practice frustrated the needs of the poorest groups and failed to improve the overall quality of life in the city. The mainstream discourse continued to insist that the housing problem is primarily a question of limited access to financial services, rather than a pattern of discrimination and neglect towards the marginalised population. Such overall urban strategies for Lima are described by Riofrío (2010) as an urban model of "housing without the city", which was imported from Chile to Peru in the process of economic neoliberalisation. The modest and fragmented reactions of the poor residents to the inconsistencies of neoliberal strategies and policies suggest a lack of political leadership and the difficulty, under the pervasiveness of market-friendly ideologies, to promote alternative responses to the long-lasting problems of metropolitan development. It has been the various kinds of popular mobilisation and the internal contradictions of hegemonic tendencies that have helped to mitigate the worst of the chaotic urban development and to favour a minimum degree of social inclusion. Paraphrasing Freud (2004), most of the people of Lima were compelled to surrender a part of their chances of happiness and political rights in exchange for the basic conditions necessary for survival in one of the most problematic Latin American megacities.

Moving back to the analytical framework (i.e. the three dimensions of the neoliberalisation discussed in the first part of this chapter), it is possible to conclude that environmental statehood changes in the 1990s were characterised by an emphasis on techno-environmental adjustments and associated monetisation initiatives (in the form of international loans and infrastructure works carried out by private companies) with less attention given to the legitimisation of neoliberalising policies (which inevitably compromised the prospects of the initially planned reforms). The persistent problems faced by those already connected to the public network and the lack of services in the newly established barriadas of Lima further damaged the perception of reforms by the general public. With the return of formal democracy in 2000, the maintenance of market-based reforms required a more convincing political justification and more effective responses to popular uneasiness. In the next decade (2001–10), the neoliberalisation of water in Lima took a more nuanced direction, with a stronger emphasis on the legitimisation of reforms, but also on renewed mechanisms to involve private sector operators. It meant a transition from techno-environmental and monetisation policies in the 1990s to more distinctive monetisation and legitimisation efforts in the 2000s. Formal democratic rule required intensive efforts to justify the direction of water management reforms, which was translated into repeated advertising campaigns by SUNASS and SEDAPAL, as well as by the newly created Vice-Ministry of Construction and Sanitation (in 2002) and the National Water Authority (in 2008). A credible political message was necessary to persuade hundreds of residents' associations of the supposed advantages of neoliberal water reforms, and to contain criticism from environmental and social NGOs and the national federation of water utility workers (FENTAP).

The second phase of water neoliberalisation in Lima started with the creation of new channels of interaction between the government and private service providers. International cooperation agencies (such as GTZ, CIDA, KFW, USAID, etc.), governmental donors and multilateral banks (European Union, JICA, OAS, World Bank) intensified their assistance in the neoliberalisation of water in Peru, by supporting governmental and non-governmental projects alike, and by searching for alternative forms of service provision. One of the first experiments was the twenty-seven-year BOT (build-operate-transfer) contract for drinking water production in the Chillón catchment (called project Blue Water or *Agua Azul*). The concession to an Italian operator was worth US$250 million and was intended to cover approximately 5 per cent of Lima's water needs. However, this very first contract has already been criticised for not favouring SEDAPAL, since only 35 per cent of the money paid to the private concessionaire is billed to the water customers due to the public utility's lack of distribution systems. Despite such evident shortcomings of these market-based solutions, the incoming president Alan García saw clear opportunities in maintaining and expanding the modernisation of the water industry in partnership with private operators. During his campaign in 2006, the phrase 'without water there is no democracy' was cleverly incorporated into García's election manifesto and, afterwards, used as a main slogan for the new government.

As an experienced politician, García evidently perceived the political advantages that could be derived from investments in the water infrastructure of the capital. At the same time García recognised that it would require additional efforts in terms of political justification to ensure popular acceptance of the renewed monetisation of water in Lima. For the business community, García seemed the ideal leader to move the agenda of the neoliberalisation of water in Peru forward. With García, the neoliberalisation of water moved from a largely economic and technocratic perspective to a more subtle coordination between economic and political goals. The advance of water neoliberalisation in Lima also benefited from the weakening of political opposition and the internal disputes between left-wing, popular sectors in the two previous decades. An unmistakable sign of the fragmentation of the workers' movement has been the collaborative attitude of the very union that directly represents the employees of SEDAPAL, called SUTESAL (moreover, a consequence of the collaborative approach of SUTESAL leaders was the growing number of contracted-out workers, who receive lower salaries than regular SEDAPAL staff, no additional benefits and no safety equipment).

In 2007, the programme Water for All (APT) was launched by the García government as one of the examples of the supposed 'Peruvian model of growth with social inclusion'. APT contained more than 300 individual projects nationwide and 150 in Lima alone, which greatly enhanced the opportunities for foreign companies (particularly American, Brazilian, Chilean and Spanish corporations) to be involved in the water services of Lima. A series of 'megaprojects' was incorporated into the APT portfolio, such as the construction of the Huachipa water treatment works and the expansion of the distribution system in the North Cone of Lima. Likewise, various PPPs were formalised in order to build a water transfer scheme from Huascacocha in the Andes (called project Marca IV), a desalination plant in the south of Lima (to be constructed by a new PPP water utility, 'Aguas del Sur de Lima'), and the sewage treatment plants of Taboada and La Chira. Overall, the initiatives included in the APT programme comprised new dams (total budget of US$480 million), systems for water potabilisation and distribution (US$787 million), wastewater treatment plants (US$468 million) and the restoration of water systems in the northern part of the city (US$570 million), according to SEDAPAL (2007a). To secure additional funds and send a message of strong commitment to neoliberal aspirations, SEDAPAL was listed on the stock market of Lima. A decree passed in June 2008, during the aforementioned 'Eighteenth Brumaire' of Alan García, authorised the water utility to negotiate at least 20 per cent of its shares.

The impact of multiple business transactions related to water through the implementation of the APT programme – made possible because of the stronger political legitimisation secured by García – went much further than infrastructure projects and large business contracts, and eventually started to permeate the everyday public perception of water issues. It is quite remarkable that some of the poorest areas of Lima, such as Pachacútec, have become the testing ground for micro-credit schemes (i.e. a form of micro-monetisation), described as the 'new paradigm' of sanitation in Peru (Baskovic, 2008). The experiment involved the creation of so-called 'small sanitation markets' and was sponsored by NGOs, government and international agencies. Local shops were encouraged to sell sanitation equipment and toilet units to the residents, with five intervening banks offering financial assistance. Credit was simplified because property deeds were not required, but only proof of employment and some evidence of property tenure (*constancia de posición*). Although on paper it may have seemed an interesting idea, in practice the promotion of micro-credit met with scepticism from the locals, as had been the case with previous initiatives promoted by international agencies and often rejected by the population. Moreover, the initiative struggled to make progress: it started with twenty-one promoters, and after a year of activity had only five, rather unenthusiastic, agents. Local residents complained that the equipment and the technology were not appropriate to their wooden houses and, ultimately, only the better off in the community could really benefit from the micro-credit conditions.

The neoliberalisation of water services since 1990 has been a staged combination of the three complementary dimensions of the neoliberalisation of the water sector, which have been carefully orchestrated in order to create a favourable business atmosphere and transfer part of the responsibility for public services to the private sector. It is possible to confirm the patchy achievements and widespread insufficiencies of water neoliberalisation in Lima. While technical solutions have failed to prevent the degradation of surface and ground water reserves, the involvement of private operators has been erratic and dependent on public funds and higher tariffs. Between 2001 and 2010, water production increased by 3.1 per cent, while during the same period the tariffs were increased by 53.8 per cent. Lima is now a city where money circulates through household water tariffs (US$350 million in 2008, according to SEDAPAL annual reports), local water vendors (there are still hundreds of water trucks in operation) and contracts with private concessionaries, but there are still persistent management problems and uncertainties about the future of its water industry. The neoliberalisation of water in Lima demonstrates that it is an intricate, non-linear process that requires constant institutional and organisational adjustments according to evolving politico-economic circumstances. Table 5.1 shows the differences between the services provided under the conventional environmental statehood paradigm, the idealised institutional model and the neoliberalisation of the water industry of Lima related to the flexible environmental statehood which was actually implemented.

A main drawback of the neoliberalising agenda was the fact that the technocratic attitudes of SEDAPAL, in its association with international construction companies, have undermined the chance to advance other low-cost alternatives based on the more active involvement of local residents. Likewise, the investments that took place both in the 1990s and in the 2000s focused on the expansion of physical infrastructure rather than on the quality and affordability of the service. There have been two main moments of substantial investments in SEDAPAL (which correspond to the monetisation dimension of water neoliberalisation), one in the mid-1990s, which was mainly dedicated to pipeline restoration and additional sources of raw water, and another, since 2005, which focused on the improvement of primary and secondary pipelines. A significant part of these investments is supposed to be recovered by realigning customer tariffs, such as the 43.8 per cent increase between July 2006 and December 2008. Even before the conclusion of APT projects, SUNASS had already approved increases in domestic water charges to fund the construction of several initiatives (i.e. 10.37 per cent for Marca II, Huachipa, Ramal Norte and Ramal Sur, and 12.31 per cent for the Taboada sewage treatment plant and an underwater sewage pipeline). One key problem is that such an approach has essentially cemented the current framework of public–private alliances, given that the investors will obviously expect to see a return on their investments coming from water tariffs in the future.

Systematic increases in water tariffs since 1995 may have enhanced the cost-recovery capacity and financial health of the utility, but have not improved the relation between SEDAPAL and the population of Lima. On the contrary, as pointed out by a community leader during our research in the city, "SEDAPAL can only really communicate with the population via the water bill". Despite the

Table 5.1 Changes in the public water sector of Lima under the conventional environmental statehood, the idealised institutional model and the actual neoliberalising process

Attributes	Public services of the developmentalist state under the conventional environmental statehood (1930s–1980s)	Neoliberalisation of the water industry associated with the search for a flexible environmental statehood (1990s–2000s)	
		Idealised	Actually implemented
Overall aims	Economic production and provision of water to the regularised neighbourhoods of Lima	Service expansion through the reduction of state interventions and the privatisation of SEDAPAL	Infrastructure modernisation through various alliances between SEDAPAL and private companies
Policy priorities	Maximisation (supply augmentation)	Efficiency (supply and demand optimisation)	Effectiveness (facilitate supply, inform demand)
Rationale of water services	Public ownership and centralised management; nominal service tariffs	Private ownership of water utilities; tariffs that secure profitable service operation	Flexible ownership and shared management; towards cost-recovery tariffs
Key disciplines behind water management	Hydrology, law and engineering	Economics, business management and engineering	Economics, public communication and engineering
Economic reasoning	Water management helps to create the basic conditions of production	Water management as reduction of state failures	Water management as reduction of state and market failures
Environmental management approaches	Living with risk; minimal environmental concerns	Delegating environmental risk management to the private sector	Coping with risk; multiple interventions to restore environmental conditions
Social consequences	Sociospatial inequality and conflicts; partial coverage of public services	Likely exacerbation of sociospatial conflicts; increased service coverage but at higher costs	Persistence of old and new conflicts; enduring inequalities in public service provision

systematic promotion of the advantages of modern water services in the mass media, many residents resent the difficulties in communicating with official agencies, resulting in protests that are mostly ineffective. The main target areas of the APT programme, such as Pachacútec, continue to suffer from regular service delays and the population constantly has to battle to obtain information about the pace and scope of construction works. The tension between the utility and its clients is manifested through the escalation of vandalism and water meter theft, which increased from 32,256 to 85,176 between 2000 and 2007, whilst the rate of metering only increased from 62.8 per cent to 70.1 per cent in the same period (cf. SEDAPAL annual bulletins). Most of the cases of vandalism happen in low-income areas, which suggest a spontaneous reaction against the attempt of the water utility to closely monitor water usage: in 2007, 23.3 per cent of the cases occurred in Comas and 21.8 per cent in Villa El Salvador (SEDAPAL, 2007b). The neoliberalisation of the water sector of Lima has also been marred by repeated evidence of corruption. In July 2009, a cabinet minister was involved in a scandal concerning the cancellation of the contract for the construction of the aforementioned Taboada treatment plant and was accused of taking bribes (apparently to the tune of US$1 million) from private companies. A few months later, León Suematsu, then president of SEDAPAL and also Vice-Minister of Construction and Sanitation, was forced to resign due to serious allegations of corruption in the construction of a new water treatment plant, which involved members of his family, politicians and private contractors.

Although substantial sums were invested in the augmentation of the pipeline infrastructure and the operational adjustments have attracted more international contractors than SEDAPAL was able to manage, much less attention has been dedicated to increasing the long-term resilience of the metropolitan water services. This has received great condemnation not only from grassroots campaigners, but also from those groups that advocate the more straightforward neoliberalisation. The managers of multilateral agencies constantly express their frustration with what they see as the slow pace of the institutional reforms in Peru, as well as their preference for more strict market policies and utility privatisation. The fundamental criticism is that, in spite of significant works under construction, the bulk of the investments continue to rely on general taxation (i.e. the national treasury). In other words, there exists a clear uneasiness among orthodox neoliberals due to the fact that the expansion of the water infrastructure in Lima is mainly being financed by the government rather than by those that directly benefit from the investments (i.e. the customers of SEDAPAL). These same critics point out that it is not clear if, in the end, the APT programme will have sufficient resources to fulfil all its ambitious targets, particularly in the light of international financial instability since 2008. A sizeable part of government funds comes from international loans, but the willingness and ability to contract loans varies between administrations, which are seen as a barrier to long-term planning. The performance of the regulatory agency SUNASS is also considered to be one of the central problems of the reform in the water sector, because of the lack of legal instruments and sustained political interferences.

In the end, in spite of the busy agenda of neoliberalising reforms in the last two decades, the organisation and performance of the water services of Lima remain highly unreliable and their future fraught with uncertainty. The water sector has become entrapped in a vicious circle of social exclusion, passive governmental responses and fresh opportunities for a new round of demagogy and populism. Because of a single-minded focus on supply augmentation, there has been limited attention to the management of water demand (something that obviously has much less electoral visibility). The emphasis on pipeline infrastructure and the failure to address the unsustainability of water reserves was condemned by SEDAPAL's former president Carlos Silvestri as a future with "less water for all". The hectic schedule of institutional reforms and infrastructure investments has been implemented against a background of spatial and sociopolitical inequalities that have characterised the urban development of Lima in recent decades. Such patterns have been maintained and even reinforced under water neoliberalisation, which has offered only short-term answers to the challenging water problems of the capital and, in the end, has mostly benefited the same business groups and elite members of society that had historically profited from state-led interventions.

The persistent and multifaceted problems of water scarcity in Lima demonstrate the interconnections between the various mechanisms of social exclusion that have composed the overall history of environmental statehood in Peru. Water scarcity cannot be understood as an isolated phenomenon, but as a process constantly reinserted in the totality of multiple urban scarcities. Instead of a purely material phenomenon, the condition of water scarcity reflects the long-term development of the capital city in relation to the rest of the country and the internal inequalities within the metropolitan area. While the old barriadas remain areas of partial integration of urban life, the new barriadas propagate the same stratified organisation of space that presupposes renewed forms of scarcity. Likewise, despite the higher sums of capital that now circulate in the city due to the introduction of neoliberal policies in the last two decades, city expansion and economic growth have in effect accelerated the social presupposition of scarcity, as is made evident by the spread of unemployment and job insecurity, the foundation of new neighbourhoods at significant distances from the city centre and the unresponsiveness of the government to grassroots demands for water and other public services.

The constant reinforcement of multiple scarcities – due to a combination of top-down strategies and the manipulation of investments and infrastructure – has become the most basic experience in the daily struggle for survival in the periphery of such vast urban areas. In the case of the Peruvian capital, both city regeneration and water management have operated within the hegemonic asymmetries that dominate the political scene and, crucially, have reinforced disparities inherited from previous historical periods. Even when low-income areas manage to secure concessions from public authorities, infrastructure and services are typically second-class. More significantly, the dialectical interplay between scarcity and abundance has been systematically used as a political device to handle expectations in the deprived areas of the capital. The deficiencies of the public water services are less the result of state failure than the convergence of powerful private interests in the organisation of urban water systems. Scarcity is instrumental for the emergence of circumstantial 'abundances', at the price of maintaining long-established, multiple scarcities. As in the past, the recent responses to water problems are centred on the appropriation of scarcity as a productive force that serves dominant interests and political agendas. In order to search for genuine responses to the mounting water problems that trouble the low-income population, these multiple scarcities need to be considered in their totality, by acknowledging the uneven advantages accrued from the production of fluid scarcities and abundances in the city.

The complexity and challenges for reforming Lima's water sector highlight the non-linear evolution of the neoliberalisation of water and the multifaceted synergies between socioecological and politico-economic processes that underpin environmental statehood reforms. In order to address such socionatural complexity, a conceptual framework was initially introduced, which provides a more comprehensive explanation of the interplay between the different dimensions of water neoliberalisation. Rather than depicting water neoliberalisation as an ideal type process, the framework suggested earlier in this chapter accounts for the advance of neoliberalism over water systems as the contingent outcome of techno-environmental improvements, the monetisation of water services and the search for political legitimacy. Those three dimensions are inherently present in the local, national and international experiences of the neoliberalisation of water, but are always manifested in different ways according to specific demands, pressures and opportunities. The first decade of water neoliberalisation in Lima was focused on technical and economic goals, due to the precarious condition of the water services in the early 1990s and the attempt by Fujimori to reinsert the country in the globalised economy via, among other strategies, the privatisation of public utilities. However, because of the questionable legitimacy of the semi-dictatorial regime, the government was forced to postpone the desired divestiture of SEDAPAL, while sustaining the flow of investments in water infrastructure. After an interim transition under Toledo, which maintained the direction of the reforms, the election of Alan García paved the way for the return of an aggressive neoliberalisation of water in Lima, albeit with more careful efforts in terms of political legitimisation behind monetising strategies (e.g. concession to private operators and higher water tariffs). As in many other South American countries, water neoliberalisation continued to expand and increasingly attract private investors and construction companies. The introduction of the programme Water for All (APT) was instrumental in consolidating several new business opportunities, although it was always justified by a discourse of social sensibility and universal services.

Despite institutional reforms and several infrastructure initiatives, the water services of Lima remain in a context of marked social inequalities (between the services provided to consolidate areas and the settlements in the periphery), with serious uncertainties about the availability of resources (due to environmental degradation) and the long-term provision of services (because of the reliance on foreign funds and private sector expertise). Notwithstanding those manifest deficiencies, the imposition of the neoliberal agenda on the water industry of Lima has so far prevented the emergence of more systematic opposition and criticism. Political containment partially derives from temporary, localised improvements in water services and partially from the ideological pressures (reinforced daily by the mass media allied to the government) that help to demotivate protest groups and resident associations. In any event, challenges and confrontation are likely to increase due to the likely failures of the neoliberalisation of water in Lima. The continuous increase in water tariffs, which are needed to maintain the involvement of private companies, constitutes a growing area of contestation. The election of a nationalist president in 2011 (Ollanta Humala, a former army officer) failed to produce a change of policies or reduce the emphasis on neoliberal solutions, because of the influence of the hegemonic politico-economic model, that is likely to persist, perhaps in even more subtle and disguised ways. Water management in Lima, and throughout Latin America for that matter, remains a highly politicised topic, precisely because the demands and rights of the vast majority of the low-income population continue to be systematically denied by the narrow, discriminatory priorities of neoliberal public policies.

On the whole, institutional reforms and infrastructure investments have been implemented against the background of spatial and sociopolitical inequalities that has historically characterised the urban development of Lima. This pattern of unevenness has been maintained under water neoliberalisation, regardless of the discourse of universalisation and better regulation. In fact, the flexibilisation of water services, adopted in order to attract private companies to the water business, has not served the basic needs of most of those living in the capital city. Despite the rhetoric about the advantages of the neoliberalisation of water, market-based solutions have privileged the better-off minority of the population of Lima, who have enjoyed improved service without having to take responsibility for the investments made by the water utility. Neoliberalism has offered only short-term answers to the challenging water problems of the capital and has mostly benefited the same business groups and elite members of society that have launched the country down the narrow road of market globalisation. For the majority of the low-income population, water remains scarce and increasingly more expensive, and there is limited opportunity to influence policymaking, particularly due to the technocratic management of SEDAPAL. Likewise, the declining availability of water reserves around Lima and in the Andes represents a lingering threat to standards of living and to the economic development of the metropolitan region.

The neoliberalisation of the water services of Lima constitutes an unambiguous illustration of the transition to more flexible forms of environmental statehood and of the adaptive, constantly evolving basis of the neoliberal state. The concrete experience of institutional and organisational reforms proved to be more complicated than mere changes to management and commercial practices. The continuity of neoliberalisation strategies required increasing levels of flexibility and responsiveness to extra-economic pressures. On the one hand, the imposition of market-based responses to the long-standing water problems of Lima was an important element of the conservative reconfiguration of the national economy and the renovation of the Peruvian State. The reform of the water industry was an invaluable opportunity to attract international companies and help to convey the message that the country was 'open for business'. On the other hand, the advance of the neoliberalising reforms faced unexpected barriers due to the biophysical characteristics of water (a resource that is unequally distributed and requires large infrastructure works) and the persistent scepticism of the population (due to the irregularity of improvements and the deficient performance of the water utility). In order to overcome those difficulties, the state had to constantly amend its own configuration (e.g. the creation of a new regulatory agency and a new vice-ministry) and reaffirm its political commitment to market-friendly reforms (e.g. more aggressive communication campaigns, new legislation on public–private partnerships, shares of the water utility sold on the stock market, etc.).

The achievements and failures of the neoliberalisation of water in both Scotland and Peru have ultimately depended on a range of politico-economic and socioecological interactions creatively mediated by the state apparatus. The next chapter will examine the contribution of Hegelian political thinking, in particular in terms of the legitimation of flexible environmental statehood in the face of widespread socionatural and politico-ecological tensions. It will also include an analysis of the criticism of the Hegelian constitutional model and the need to go beyond the ideological basis of the contemporary (capitalist) state.

Bibliography

Alcázar, L., Xu, L.C. and Zuluaga, A.M. 2000. *Institutions, Politics, and Contracts: The Attempt to Privatize the Water and Sanitation Utility in Lima, Peru.* Policy Research Working Paper No. 2478. World Bank: Washington, DC.

Bakker, K. 2005. Neoliberalizing Nature? Market Environmentalism in Water Supply in England and Wales. *Annals of the Association of American Geographers*, 95(3), 542–65.

Bakker, K. 2010. The limits of 'neoliberal natures': Debating green neoliberalism. *Progress in Human Geography*, 34(6), 715–35.

Bernstein, S. 2001. *The Compromise of Liberal Environmentalism.* Columbia University Press: New York.

Budds, J. and McGranahan, J. 2003. Are the Debates on Water Privatization Missing the Point? Experiences from Africa, Asia and Latin America. *Environment and Urbanization*, 15(2), 87–113.

Calderón Cockburn, J. 2005. *La Ciudad Ilegal: Lima en el Siglo XX.* UNMSM, Lima.

Carr, E.R., Wingard, P.M., Yorty, S.C., Thompson, M.C., Jensen, N.K. and Roberson, J. 2007. Applying DPSIR to Sustainable Development. *International Journal of Sustainable Development*, 14(6), 543–55.

Castree, N. 2008a. Neoliberalising Nature: The Logics of Deregulation and Reregulation. *Environment and Planning A*, 40(1), 131–52.

Castree, N. 2008b. Neoliberalising Nature: Processes, Effects, and Evaluations. *Environment and Planning A*, 40(1), 153–73.

Cohen, B. 2004. Urban Growth in Developing Countries: A Review of Current Trends and a Caution Regarding Existing Forecasts. *World Development*, 32(1), 23–51.

De Soto, H. 1990. *El Otro Sendero*. 9th edition. Instituto Liberdad y Democracia: Lima.

European Commission. 2000. *Water Framework Directive*. Directive 2000/60/EC of the European Parliament and of the Council of 23 October 2000 Establishing a Framework for Community Action in the Field of Water Policy.

Fine, B. and Saad-Filho, A. 2010. *Marx's Capital*. 5th edition. Pluto Press: London.

Fiorino, D.J. 2006. *New Environmental Regulation*. MIT Press: Cambridge, MA.

Freud, S. 2004. *Civilization and its Discontents*. Trans. D. McLintock. Penguin: London.

Gabrielsen, P. and Bosch, P. 2003. *Environmental Indicators: Typology and Use in Reporting*. European Environmental Agency: Copenhagen.

Habermas, J. 1991. *The Structural Transformations of the Public Sphere: An Inquiry into a Category of Bourgeois Society*. Trans. T. Burger. MIT Press: Cambridge, MA.

Heynen, N. and Robbins, P. 2005. The Neoliberalisation of Nature: Governance, Privatization, Enclosure and Valuation. *Capitalism Nature Socialism*, 16(1), 5–8.

Ioris, A.A.R. 2012. The Political Geography of Environmental Regulation: Implementing the Water Framework Directive in the Douro River Basin, Portugal. *Scottish Geographical Journal*, 128(1), 1–23.

Ison, R. and Watson, D. 2007. Illuminating the Possibilities for Social Learning in the Management of Scotland's Water. *Ecology and Society*, 12(1), 21.

Jordan, A. 2000. The Politics of Multilevel Environmental Governance: Subsidiarity and Environmental Policy in the European Union. *Environment and Planning A*, 32(7), 1307–24.

Kaika, M. and Page, B. 2003. The EU Water Framework Directive: Part 1. European Policy-making and the Changing Topography of Lobbying. *European Environment*, 13(6), 314–27.

Kanakoudis, V. and Tsitsifli, S. 2010. Ongoing Evaluation of the WFD 2000/60/EC Implementation Process in the European Union, Seven Years after its Launch: Are We Behind Schedule? *Water Policy*, 12(1), 20–31.

Kant, I. 1929 [1781]. *Immanuel Kant's Critique of Pure Reason*. Trans. N.K. Smith. Macmillan: Basingstoke and London.

Kastens, B. and Newig, J. 2007. The Water Framework Directive and Agricultural Nitrate Pollution: Will Great Expectations in Brussels be Dashed in Lower Saxony? *European Environment*, 17(4), 231–46.

Keulartz, J. and Leistra. G. (eds). 2008. *Legitimacy in European Nature Conservation Policy: Case Studies in Multilevel Governance*. Springer: Dordrecht and London.

Kirk, E.A., Reeves, A.D. and Blackstock, K.L. 2007. Path Dependency and the Implementation of Environmental Regulation. *Environment and Planning C*, 25(2), 250–68.

Liefferink, D., Wiering, M. and Leroy, P. 2011. The Water Framework Directive: Redesigning the Map of Europe? In: *A History of Water, Series II, Vol. 3: Water Geopolitics and the New World Order*, Tvedt, T., Chapman, G. and Hagen, G. (eds). I.B. Taurus: London/New York, pp. 241–60.

Marx, K. 1973 [1857–8]. *Grundrisse*. Trans. M. Nicolaus. Penguin and New Left Review: London.

McAfee, K. and Shapiro, E.N. 2010. Payment for Ecosystem Services in Mexico: Nature, Neoliberalism, Social Movements, and the State. *Annals of the Association of American Geographers*, 100(3), 579–99.

McCulloch, C.S. and Ioris, A.A.R. 2007. Putting Politics into IWRM. *Geophysical Research Abstracts*, Vol. 9, 02981.

Moran, D., MacLeod, M., McVittie, A., Lago, M. and Oglethorpe, D. 2007. Dynamics of Water Use in Scotland. *Water and Environment Journal*, 21(4), 241–51.

Murakami, Y. 2008. Política Peruana después de Fujimori: Fragmentación Política y poca Institucionalización. *CIAS Discussion Paper*, 5, 41–63.

Niemeijer, D. and de Groot, R. 2008. Framing Environmental Indicators: Moving from Causal Chains to Causal Networks. *Environment, Development and Sustainability*, 10(1), 89–106.

Picciotto, S. 2011. International Transformations of the Capitalist State. *Antipode*, 43(1), 87–107.

Riofrío, G. 2010. Alan García, Alcalde de Lima. In: *Perú Hoy. Desarrollo, Democracia y otras Fantasías*. DESCO: Lima, pp. 71–96.

Sangameswaran, P. 2009. Neoliberalism and Water Reforms in Western India: Commercialization, Self-sufficiency, and Regulatory Bodies. *Geoforum*, 40(2), 228–38.

SASE. 2002. *Estudio sobre la Dinámica de los Asentamientos Humanos*. PDPU/COFOPRI: Lima.

Scottish Government. 2008. *Implementation of the Water Environment and Water Services (Scotland) Act 2003: Annual Report to the Scottish Parliament – 2007*. Scottish Government: Edinburgh.

Scottish Water. 2007. *Annual Report and Accounts 2006/2007*. Scottish Water: Dunfermline.

SEDAPAL. 2005. *Plan Maestro Optimizado*. SEDAPAL: Lima.

SEDAPAL. 2007a. *Plans for the Water and Wastewater System in Lima: 2008–2015*. SEDAPAL: Lima.

SEDAPAL. 2007b. *Informe de Gestión Financiera y Presupuestaria*. SEDAPAL: Lima.

SEPA. 2005. *Solway Tweed River Basin District*. Scottish Environment Protection Agency and Environment Agency. Stirling.

Slater, D. 1989. *Territory and State Power in Latin America: The Peruvian Case*. St Martin's Press: New York.

Smith, L. 2004. The Murky Waters of the Second Wave of Neoliberalism: Corporatization as a Service Delivery Model in Cape Town. *Geoforum*, 35(3), 375–93.

Thompson, N. 2006. The Practice of Government in a Devolved Scotland: The Case of the Designation of the Cairngorms National Park. *Environment and Planning C*, 24(3), 459–72.

Ugarteche, O. 1999. *La Arqueología de la Modernidad*. DESCO: Lima.

Vargas Llosa, M. 2005. *El Pez en el Agua*. Alfaguara: Madrid.

Verdera, F. 2007. *La Pobreza en el Perú*. IEP: Lima.

Watson, N., Deeming, H. and Treffny, R. 2009. Beyond Bureaucracy? Assessing Institutional Change in the Governance of Water in England and Wales. *Water Alternatives*, 2(3), 448–60.

WIC. 2007. *Strategic Review of Charges 2010–14: Methodology. Volume 1: Financing and Governance of Scottish Water*. Water Industry Commission for Scotland: Stirling.

Zito, A.R. 2000. *Creating Environmental Policy in the European Union*. Palgrave Macmillan: Basingstoke.

6 Flexible environmental statehood as the realisation (actualisation) of the Hegelian state

Hegel and statehood

As discussed in the previous chapters, the contemporary (capitalist) state underwent significant transformations in the first half of the twentieth century in order to respond to mounting environmental degradation and to better coordinate access to territorial resources (e.g. coal, oil, minerals, water, timber, etc.). In most countries, a dedicated branch of the state apparatus was put in charge of the new area of environmental statehood, which was implemented through the introduction of ever more complex legislation and techno-bureaucratic agencies. The executive and more tangible side of environmental statehood included a range of transient institutional arrangements and regulatory mechanisms that constitute the state-fix of each specific historical period. After the Second World War, the organisation of environmental statehood was intensified as part of the increasingly interventionist agenda of the welfare or welfare-developmentalist state. Starting in the Western countries, considerable environmental policy-making organisations were instituted by most national governments, together with agencies responsible for national parks, environmental impact assessments, licences and permits, etc. This model of conventional environmental statehood represented the first attempt to systematically incorporate an environmental agenda into the daily activities of the state. However, it soon started to reveal a series of weaknesses and shortcomings that were criticised by both allies and enemies of the national governments. The main critique was that, despite advances in some areas, conventional environmental statehood largely failed to contain an escalating amount of socioecological impacts caused by economic development and state-led agroindustrial expansion. Conventional environmental statehood was also seen as too slow, with responses formulated mostly after the processes of environmental disruption were already in place. From a politico-ecological perspective, conventional environmental statehood was an institutional compromise that operated as an adjunct to stronger political and economic priorities, that is to say, it was never meant to question the legitimacy of the state and the prevailing trend of production, consumption and wastage.

Due to its operational and political inadequacies, the conventional model of environmental statehood was (partially) replaced by the more flexible and responsive approaches introduced in the last few decades of the century. This more recent, and still largely dominant, framework was again not intended to alter the anti-commons and privatist pillars of mainstream public policies (including environmental policies that became increasingly mediated by economic and monetary considerations). The ultimate intention of institutional flexibilisation reforms was to simplify environmental regulation, remove over-bureaucracy and stimulate economic activities around environmental conservation. The emphasis shifted from environmental restrictions to environmental governance through the substitution of previous command-and-control processes with more interactive and adaptive procedures. This renovation of environmental statehood was directly associated with the need to address the multiple crises of capital accumulation and the inability of the Keynesian state to protect the average rates of profit (discussed in Harvey, 2005). There was, therefore, a contingent, but highly significant, relationship between the transition to flexible environmental statehood and the adoption of wider neoliberal policies (including a range of sectoral strategies aimed at neoliberalising socionature). The reconfiguration of the state apparatus under neoliberal pressures produced a unique opportunity for the reorientation of the logic and practice of environmental management in order to both address the institutional deficiencies or rigidities of the earlier statehood model and to create new opportunities for the circulation and accumulation of capital. The main outcome was the reinsertion of environmental issues into the state–society–nature trialectics following the priorities of economic globalisation and the deepening of market-based public policies. This resulted in the extraction of profit from not only the exploitation of natural resources, but also the management and conservation of socionature itself.

In practical terms, flexible environmental statehood has been simultaneously neoliberal and 'more-than-neoliberal', in the sense that it has also retained many features of the conventional model of environmental statehood. Somewhat paradoxically, flexible environmental statehood has entailed the escalation, instead of the proclaimed decline, of state interventions. This is demonstrated, for example, by the rising number of state agencies and environmental policies that have increased the opportunities for the state to interfere. Furthermore, the introduction of flexible environmental statehood has been a puzzling experience that combines, on the one hand, a larger number of social actors and more integrated approaches, and, on the other hand, the persistence and even intensification of multiple processes of environmental degradation. Expressions such as sustainable development, public participation and environmental governance have populated the pages of policy documents and legislation, but have also served to legitimise the adoption of controlled adjustments of environmental regulation and facilitated the acceptance of market-based environmental management strategies (for example, carbon trade, water markets and the payment for ecosystem services). Although this intriguing process of change has been the object of broad academic debate, the examination of the politico-philosophical basis of more flexible environmental statehood has been largely insufficient and often superficial. The purpose of this chapter is to understand the consolidation of flexible environmental statehood as an integral element of the state apparatus in the early twenty-first century. In order to achieve this objective, it will be necessary to analyse the tradition of Western political philosophy informing state reforms and, in particular, the rationalisation and legitimation of flexible environmental statehood influenced by Hegelian political thinking.

The recent changes in environmental statehood represent a shift from the Hobbesian basis of power and authority, and the Kantian ideas about liberalism and political transition, to the shrewdness of Hegelian state theories and propositions. Although the new model of environmental statehood offers only limited responses to environmental impacts and conflicts associated with conventional environmental statehood, it has been portrayed by policy-makers as the definitive expression of wisdom, democracy and scientific aptness of the state apparatus for dealing with socioecological issues. The agenda of flexible environmental statehood bears a close resemblance to Hegel's plans for the constitutional state in charge of growing social inequalities and persistent political segregation in early nineteenth-century Prussia. Some may find it surprising that the foundation and legitimacy of flexible environmental statehood had an early advocacy in the political writings of Hegel. Nonetheless, it should be remembered that Hegel has played a very important role as one of the most creative, and ambitious, philosophers of modern Europe. According to Habermas, Hegel was the first thinker for whom modernity was a philosophical problem and, as a result, Hegel became the first philosopher of modernity (mentioned in Rockmore, 1989). Hegel insisted on the emancipation of society and politics from religion (whilst maintaining a highly idealised ontology) and encouraged the unification of tradition and modernity. Beyond the liberalism of Locke, the liberal-utilitarianism of Mill and Bentham, and the rationalisation of Weber, the political and moral claims of Hegel provided the intellectual tools for the consolidation of the framework of environmental statehood that prevails around the world nowadays, particularly in the European Union.

To prove the relevance of Hegel to the twenty-first century political ecology of the state, it is necessary to examine the intricacies and the historico-geographical context of Hegelian political philosophy. In Hegel's lifetime, Prussia was in a peculiar situation with a still incomplete bourgeois revolution and a bourgeoisie that could not rise against the government (personified by the king) without itself being bitten in the rear by the proletariat. Notwithstanding this situation, the Prussian State needed to be rearranged and gradually transformed to fulfil many new roles in a continent increasingly dominated by capitalist relations of production. The present of the state is incomplete without the conscious understanding of its past, but the action of the state is supposed to create history itself (Hegel, 1953). Hegel was an acute observer of the problems and demands of the existing state system, and attempted to cleverly accommodate the needs of the declining aristocracy and the aspirations of the rising bourgeoisie. He dealt with the perceived weaknesses of the many German states not only in philosophical texts, but also in his journalistic and political writings.

In particular in *The German Constitution*, a manuscript prepared around 1800 and left unpublished, Hegel provides an astute assessment of the fragmentation and authoritarianism of the German peoples, in particular the separation between theory (the institutions of the empire) and practice (the group of independent political units). For Hegel, one of the main deficiencies of the Prussian State of his time was the lack of a strong, committed leadership. "A multitude of human beings can only call itself a state if it be united for the common defence of the entirety of its property" (Hegel, 1964: 153).

Inspired by Machiavelli's elaboration on pragmatism and authority, Hegel saw the successful state as the ultimate 'prince' of the modern world. In contrast to Machiavelli, however, according to Hegel the state should be guided by the principles of justice and the search for rationality and effectiveness. A legitimised authority and absolute knowledge were supposed to be achieved by the state by virtue of logic and dialectics. The Hegelian state is basically the result of a desired balance between force and wisdom, the guardian of a contained form of democracy, but also the legitimate upholder of multiple forms of social and geographical inequalities (such as wealth, fiscal rights and duties).

> We also regard that people as fortunate to which [sic] the state gives a free hand in subordinate general activities, just as we regard a public authority as infinitely strong if it can be supported by the free and unregimented spirit of its people.
>
> (Hegel, 1964: 164)

Hegel was not only one of the main interpreters of a fast-changing reality, but his extensive, highly unique work sought to address several crucial problems in the organisation of the Prussian and other European states. In that process, Hegel combined the medieval origins of the modern state with the emerging nationalistic and economic demands of an evolving capitalist society (Lee, 2008). The philosopher recognised the responsibilities of the state as part of its mission to promote what is 'right', in other words, what is supposed to derive from the rational search for a higher moral ground. According to Hegel's political system, the state should become an institutional centre of normative, social and economic driving-forces.

The Hegelian state was portrayed as a congregation of estates and corporations – representing the bourgeoisie, the aristocracy and the peasantry – that should operate primarily as an ethical community. Instead of an entity based on absolutist principles, Hegel attributed to the state – which he described as the complex association between the monarch, the assembly of estates and the executive branches of government – the elevated mission of reconciling personal wants with the embodiment of universal ends (Hegel, 2008). Hegel didn't see the state as primarily an instrument to safeguard people's self-interest (which he located in civil society, where personal aims are mediated by the needs of others), but as the guardian of an ethical life and universal altruism (Avineri, 1972). In this respect he differed from authors working in the same period, such as Fichte, who perceived state functions as solely to protect the rights of individuals. Although many critics minimise Hegel's early concerns about the pauperising effects of industrialisation and the harmful consequences of capitalist production, the philosopher closely followed the perceived impoverishment of agriculture and industry workers throughout his career. His analysis of the reform of the electoral system in England, which aimed to improve parliamentarian representation and expand electoral rights, was considered insufficient as a means to address the condition of the destitute population (Dickey and Nisbet, 1999).

Despite Hegel's calls for (bourgeois) democracy and development, his reasoning was attacked immediately after his death by philosophers such as Kierkegaard, Schelling and Feuerbach. The philosopher was condemned for several alleged mistakes, such as the apology for Prussian power, the glorification of war, the end of history and the denial of the law of contradiction (Stewart, 1996). A century later, Hegel was again criticised for offering philosophical excuses for authoritarian forms of government, such as in Nazi Germany (Popper, 2002) and for promoting the virtues of absolute monarchy (Russell, 1950). A more careful reading, however, demonstrates that these forms of criticism are, by and large, reductionist misrepresentations of Hegel's sophisticated, but misleading, political elaboration. Probably the two main features of Hegel's political thinking were the separation of state and civil society (Pelczynski, 1984) and the incorporation of history within a philosophically relevant system, rather than a sequence of accidental and arbitrary events (Avineri, 1968). The limits and the legitimacy of state power remains a controversial question, which has persisted from Machiavelli to Franz Fanon and many other authors, but the ambitious ideas of Hegel continue to occupy centre stage. Hegel's state theorisation proved to be more subtle and complex – even if highly contradictory, as shown by Marx – than most of the comparable political theories. The task now is to present and interpret Hegel's political system and assess how it has informed the evolution of environmental statehood.

The nuances of Hegel's state model

Hegel's state system only makes sense as a pursuit of the perfect public service and the moral state if these are to be pursued through an ethical life. According to Hegel, the whole ethical basis of the state emanates from the Idea, the spirit, the absolute essence. Here we can see an attempt to restore the Platonic conceptualisation of the state as a system governed by those who excel in philosophy and military art, which includes unpaid officers of great authority and moral rectitude (Book VIII of The Republic). Plato famously listed a series of 'imperfect' states that are like men, because they are made of men; for instance, the accumulation of wealth by men leads to the proliferation of vices and reduction of virtues, which have serious repercussions on the state. There is, therefore, a need to overcome the 'imperfect states' – which for Plato not only consist of oligarchic regimes, but also include tyrannical and democratic political systems – through moral education. Hegel claimed to resolve the fundamental tension between the public and private interests, present in the Platonic tradition, with a 'mediation of the will' between the family, civil society and the state. These are the three moments of ethical life, with the family as the realm of particular altruism, civil society as the realm of universal egoism (self-interest and economic transactions) and the state as the realm of universal altruism and solidarity (Hegel, 2008: 161–2). Following Plato, the Hegelian paradigm of public administration comprises a higher state level consisting of the monarch, parliament and government, and an inferior state level consisting of the judiciary and general public administration (Bobbio, 1995). The state is thus portrayed as a system of integration aimed at overcoming individualism in the economic life and the centrifugal forces of the market (Avineri, 1972). The Hegelian defence of the constitutional monarchy was an attempt to reconcile in the figure of the ruler both institutional stability and the expression of the will of the people (Levin and Williams, 1987).

Hegel's framework was clearly superior to Weber's ideal type of state based on moral principles and a rigid rationality (i.e. for Weber the rationalisation of Western civilisation is, to a large extent, the projection of the state's own rationality). Weber's model of ethical rationalisation and institutional embodiment of consciousness is too inflexible to demonstrate, in a convincing way, the interplay between state, economy and the rest of socionature. While Weber studied the intricacies of organised bureaucracies and warned against the risk of an oligarchic capture of the state, his conceptual and historical interpretation reveals a great deal of pessimism and could be easily appropriated for the justification of overly centralised forms of public administration. In the case of Hegel, reason is an elevated category that allows the state to operate together, and in a more productive way, with the most influential groups of society. Hegel claims that reason and morality are actualised in the very structures of the social world. For Hegel, reason (the apprehension of dialectical differences) is more than simple understanding (distinguishing one thing from another). By the same token, a judgment is different to a simple comprehension because it is "the proximate *realization* of the Notion" (Hegel, 1969: 623), that is, cognition derives from the unity of theory and practice. Hegel is particularly critical of Kant's attempt to derive concrete moral judgments from the commitment to practical reason alone (West, 2012). According to Hegel, as had been previously shown in the case of Rousseau's dissatisfaction with the technological corruption of morality and politics, ethical life can only be sustained through involvement in a concrete community. The biological inclinations of the individual are thus not incompatible with the moral or rational ones (as Kant claimed), but morality should be the realisation of such inclinations. The individual (the citizen) is seen as the product, rather than the premise of social order (West, 2012). In that sense, Hegel portrays and promotes the state as an ethical institution able to give rational expression to certain progressive tendencies seen as immanent in politics (Smith, 1989).[1]

The political and constitutional writings of Hegel are primarily concerned with ethics and aim to discuss how people and states should behave and how they could be rationally judged (Knowles, 2002). In his main political text,

Outlines of the Philosophy of Right – essentially, a compilation of lectures and notes taken or expanded by his disciples – Hegel indicated that his ultimate endeavour was to *"apprehend and present the state as something inherently rational"* (Hegel, 2008: 15). In the preface to this book, Hegel unleashes his famous claim that "what is rational is actual and what is actual is rational", which is presented as an affirmation that reason is an actual power in the world (moreover, this equality between the rational and the actual attracts conflicting interpretations and was very controversial even in Hegel's own time, as examined by Stewart, 1996). Following this association between rational and actual, Hegel develops his peculiar political conceptualisation in the form of an emphasis on the coincidence between world history and rationality, which designates the state as the guarantor of the systematic actualisation of reason. The realisation ('actualisation' in the Hegelian vocabulary) of reason is the fundamental purpose of the state, as the conveyor of the perfect social order and the conciliator of multiple conflicts of interest. For Hegel, history evolves through dialectics towards the attainment of the 'Absolute', that is, the self-reflective appropriation of the whole development process. Because the state is 'inherently rational' it becomes the main promoter and defender of reason as the main, legitimate force in the course of historical change.

It is important to appreciate Hegel's terminology in order to properly unveil his contribution to political debate about the state. In the *Science of Logic* Hegel offered the definition "Actuality is the unity of essence with Existence" (Knowles, 2002: 68), but the real and the actual don't necessarily coincide, because the real can, and must, be improved. This improvement is part of the actualisation of reason and the perfecting of reality. The effort to improve the real means unifying essence with existence, that is, the enrichment of rationality. The subtlety of Hegel in his advocacy of authority and political legitimisation stood out from the rigidity or utopianism of other authors of the same period. One such author was Humboldt, the precursor of the liberalism of J.S. Mill, who saw the main task of the state as promoting happiness and preventing evil

> *The solicitude of a State for the positive welfare of its citizens must further be harmful, in that it has to operate upon a promiscuous mass of individualities, and therefore does harm to these by measures which cannot meet individual cases.*

Humboldt, 1993: 27

Knowles (2002) observes that, by claiming that the rational is actual, Hegel combines the organisational potential of the social world with the ethical demands on the individual. Following this line of reasoning, individual freedom and social rights presuppose social norms and ethical conditions, that is, freedom becomes the manifestation of the rational. The Hegelian goal is essentially the affirmation of rational moral obligations and the construction of a society that, because of its self-proclaimed success, renounces the need for a dramatic transformation (such as the violence of the French Revolution that shocked Hegel profoundly).

The originality of Hegel's insights apparently provide the conditions for the guarantee of individual freedom in accordance with the authority of the state; that is to say that ethical life becomes the entailment of a higher form of rationality (i.e. the actual) and the perfect coordination between state, society and the individual. While Hegel's idealism retained some elements of the Romanticism of his time (which he tried to avoid, but not completely), his philosophy intended to deal with the concrete, dynamic totality of the world (Toews, 1993). In his view, the "state is rational in and for itself inasmuch as it is the actuality of the substantial will which it possesses in the particular self-consciousness that has been raised to its universality" (Hegel, 2008: 228). Religion is also an important element of the attainment of universal reason and plays a key role in political action as the entry point, and the definitive pillar, of an ethical civil life rationalised by the state. According to his constitutional plan, Hegel anticipated the solution to the crucial problem of how the state should administer growing socioeconomic complexity and rising political pressures associated with technological, commercial and cultural integration in the globalised world. By situating the state at the ethical crossroads between the universal and the individual, between collective needs and private gain, Hegel produced a very appealing model for the state as "the actuality of concrete freedom" (Hegel, 2008: 235). The state is not the mere instrument of particular individuals and narrow ends, but it is an organic, ethical creature in which the totality presupposes its parts (Knowles, 2002). The state is the powerful, but ethical, entity that conducts the whole social fabric towards the kingdom of reason. More importantly, the dialectics between the imperfect real (full of potentialities) and the desirable actual (the full expression of rationality), paved the way for the organisation of civil society under the framework of the advanced capitalist economy. Much beyond the crude liberalism of the nineteenth century, Hegel already described the state as a rational organism that presupposes and expresses the life of civil society, the individual and, crucially, the economy. The separation of the identification of the individual with the collective and his or her participation in the state is something that Hegel calls alienation, and this should be avoided (Plant, 1973). In the *Philosophy of History*, Hegel even argues that the modern state is inseparable from the fulfilment of the 'general theory of ethical life', which happens when the individual participates in the collective social enterprise.

Besides the ethical presuppositions of the Hegelian state, the ingenious analytical method espoused by Hegel (which was praised by generations of followers, including Adorno, Marcuse, Sartre, Lefebvre, Rawls, among many others) had the advantage of bringing philosophy to the centre of interpersonal relations and to the core responsibilities of the state. The *Outlines of the Philosophy of Right* was not a purely speculative book, it "was intended as a contribution to practical philosophy", as a tool to evaluate and direct action (praxis) and to overcome the contradictions within the liberal state (Smith, 1989: 136). This particular text can "be described as a philosophical reconstruction of modern ethical life (*Sittlichkeit*) – the totality of ideas, practices, sentiments and relations which not only prevail in fact, but are regarded by the modern man as valid in some normative sense" (Pelczynski, 1984: 7). While Hegel remained a classical idealist philosopher – to the extent that he argued that for "the idea of the state one must not look to particular states or particular institutions; rather, the idea, this real God, must be contemplated by itself" (*Philosophy of Law*, quoted in Adorno, 1973: 335) – his interpretation of the span of developments includes an immanent encounter between emancipatory reason and historical circumstances. For the followers of Hegel, here resides one of the main contributions of the great philosopher, the notion of the 'self' or the 'subject' as the outcome of movement and constant displacement. It is the dialectics between superseding and preserving, the constant encounter with the other that changes the one. Concrete, actualised rationality is never secure in its existence; for Hegel it is always contaminated by the opposite in a perpetual process of dislocation (Gidwani, 2008).

Nevertheless, the moral elements of Hegelian philosophy have been challenged by many authors due to the representation of Hegel as a historical determinist and, more recently, the criticism offered by post-modern political philosophers and post-structuralist thinkers. The Hegelian politico-philosophical system has proved extremely controversial and can easily lead to different interpretations. His method of inquiry – basically, the dialectics of affirmation through negation – allowed for an innovative analysis of constitutional matters, but the conclusions were not easily applied to situations where reason and ethics were secondary aims. Weber, for instance, criticised Hegel's teleological construction of history and, as a response, put forward a distinctively non-Hegelian epistemological and methodological approach (which, for Sager and Rosser, 2009, nonetheless had some surprising connections with the Hegelian framework Weber so eagerly wished to rebuke). Likewise, more recent authors (e.g. Levinas, Deluze, Derrida, Lyotard) have a major problem with Hegel in his supreme incarnation of modernity and modern rationality, which leaves no room for social differences beyond pre-established categories.

In an attempt to rescue Hegel from such criticism, Žižek (2011) claims that the Hegelian notions of totality and historical necessity are, in effect, elements of a flexible reasoning that imply a radically open contingency of history. The relation between contingency and necessity is dialectical, in the sense that there is a necessity for contingencies and, more radically, a contingency of necessities (i.e. things become necessary only in a contingent way). The relation between past and present is also dialectical, where the present is obviously influenced by the past, but the past is also reinterpreted and reconstructed by the present. Rather than the trends of history being determined a priori by some overpowering force, historical necessities can only be explained retrospectively. Therefore, the Hegelian necessity should be seen not as a cause, but as the central property of the process of change (Mann, 2008). Nancy (1997: 105) emphasises, in his study of the Hegelian negation, that "*La liberté et la négativité s'exposent ainsi mutuellement*". For Nancy, negation becomes the locus of creative criticism, the force of historico-geographical necessity. The Real (with a capital 'R') is simultaneously the thing "to which direct access is not possible and the obstacle that prevents this direct access" (Žižek, 2003: 77). Following the Hegelian conceptualisation of dialectics, the Real becomes a parallax, that is, "it is just a gap between two points of perspective, perceptible only in the shift from the one to the other" (Žižek, 2006: 26). The notions of historical necessity and dialectics are particularly relevant to an understanding of the Hegelian theorisation of the present-day state apparatus, which henceforth becomes necessary (in the sense of a retrospective, contingent explanation) for the dialectical realisation of reason.

For his followers, the Hegelian philosophical system equipped the state with a defensible reconciliation between idealism and religion, morality and ethical life, and universal notions and historical contingencies. As a result, the state is supposedly predicated upon the development of relations of production and liberal politics, but it also begins to represent the culmination of multiple, expanding contingencies. In his own time, Hegel was (prudently) critical of the Prussian State – as far as this was possible under an authoritarian regime – but he also saw great potential in its 'rationalised' development. The emerging bourgeois state was seen by Hegel as a necessity of capitalism, but also as the best assurance that the expansion and prolongation of the qualities of capitalism could be achieved. State power was no longer conceived as only the authoritarian instrument of the dominant groups, but according to Hegel it should be involved in the complexities of wealth creation and accumulation, which necessarily requires the rule of law and ingenious forms of ideology:

> these simple *thoughts* of Good and Bad are likewise immediately self-alienated; they are *actual* and are present in actual consciousness as *objective* moments. Thus the first essence is *state power*, the other is *wealth*. As state power is the simple *substance*, so too is the universal '*work*' – the absolute 'heart of the matter' itself in which individuals find their *essential* nature expressed, and where their separate individuality is merely a consciousness of their *universality*.
>
> (Hegel, 1977: 301)

In a world that was still getting rid of absolutist monarchies using the vacillating advances promoted by the bourgeoisie, Hegel advanced a sophisticated critique of the state that was able to reconcile renovation and permanence, rupture and legitimacy, and partial political inclusion and socioeconomic stratification. Hegel's work is superior to other comparable theories that typically described the state as being above society (epitomised by the rights of the sovereign). In a situation like that of Germany and Europe in the post-Napoleonic period, who were still struggling to contain the ambitions of the restored royal families, Hegel's ideas about a rational (i.e. effective) state in charge of regulating market development and economic inequalities provided strong backing for the growth of a new constitutional order that was instrumental in the expansion of civil liberties akin to the expanding capitalist relations of production. Moreover, the politico-philosophical writings of Hegel only had a secondary impact on the Prussian Restoration. Later, a more nationalistic and top-down form of government was put in place by Bismarck for the formation of the German Empire instead of the moderate liberal positions sponsored by Hegel.

It took almost two centuries, after the collapses of successive aristocratic, liberal, fascist and Keynesian state formations, for the Hegelian political model to be effectively realised, particularly by the European Union with its constant attempts to contain, through complex regulation and large-scale compromise, the contradictions of a post-industrial capitalist society. It is, therefore, no surprise that Hegel has been considered "*le penseur inaugural du monde contemporain*" (Nancy, 1997: 5) and, for the current generation of his followers, "the time of Hegel still lies ahead – Hegel's century will be the twenty-first" (Žižek, 2011: xi). The main innovation of Hegel's political argument was possibly the conceptualisation of the state as an adaptive, flexible entity with more functions than just the bare affirmation of power and the most immediate demands of the stronger social groups. Ahead of his time, Hegel was already concerned with some of the core problems of late modernity, in particular, how to integrate society and share social values when the ideology that prevails is markedly individualistic, pluralistic and centrifugal (O'Hagan, 1987). Crucially, the opportunity to formulate Hegelian-informed responses to these social and political questions coincided with the reorganisation of environmental statehood and the search for new patterns of political legitimacy. In that sense, the Hegelian argument in favour of a more responsive state apparatus offered valuable assistance to the reorganisation of environmental statehood and to the incorporation of the discourse of governance, modernisation and sustainability.[2]

Hegel notably broadened the claims made by David Hume (in the *First Principles of Government*) that all governments (i.e. "the authority of the few over the many") are founded upon interest, power and property. According to Hegel, the state is a more intricate organisation than the one conceived by Hume, because it is also the holder of truth and the expression of the universal will. "The originality of Hegel's political philosophy, compared with that of other modern thinkers, consists in grounding it not in some universal characteristics of human nature or in the idea of fundamental human rights, but in ethical life" (Pelczynski, 1984: 7–8). The two mechanisms that maintain the various spheres and estates of society should be the figure of the monarch and, at the personal level, the education of citizens for the fulfilment of an ethical life. In order for that plan to work, the state requires a civil society and people who are willing to resolve disputes through civil engagement in the Assembly of Estates and who are also agreeable to the overarching social institutions of an emerging capitalist society. Hegel (2008) emphasised the essential separation between the sphere of 'private affairs' (the family and, as a separate realm, the economic activities of civil society) and the sphere of 'public affairs' (the state). According to Hegel, the subject is necessarily embodied in social relations, and this embodiment is both the condition of his or her existence and the expression of what he or she is. The subsumption, the full interiorisation of individuals and their social relations to the state, is a precondition, and also a consequence, of the perfect functioning of the commonwealth (i.e. the state).

The state idealised by Hegel is instituted upon ethics, reason and common understandings. It needs to be rational, and it is precisely this rationality that serves as justification of political authority and, when necessary, the use of force. Through his elaboration on the ethical and rational state, Hegel tackled some of the main political challenges of the time, that is to say, how to reconcile religious thoughts and stratification with the economic and political demands of the rising capitalist society. Hegel's solution was to extract dialectics from both the Gospels and humanised religion, with the final result being a system that is internally coherent and able to extend in time and space. Playing with words (for example, in his *Lectures on Philosophy of Religion*, "What God creates, God himself is"), Hegel provides a contingent, but powerful, association between God, the Idea and his teleological claims. Hegel was thus the best translator and advocate of a constitutional order beyond the rigid authoritarianism of Hobbes, the naturalistic forms of government of Montesquieu and the more abstract rationalisation of Kant. Any opposition between citizens and the state is, in theory, overcome through Hegel's defence of the embodiment in the state apparatus of the universe of human life and the incorporation of the individual by the state (Taylor, 1979). More importantly here, Hegel's oblique, at times indecisive, philosophical categories furnished the state with the ability to be both reactionary and transformative (something which would later prove very appropriate for dealing with environmental problems).

Hegel's defence of the state is directly associated with human reason and desire for freedom, but it is also the reflection of the Idea, that is, the state as the phenomenon of the Idea. The Idea is the a priori truth and does not depend on the actual existence of things. The concept of the object is enough to know about the object, to be precise, empirical understanding is dispensed in favour of conceptualisation. "History is spirit [Idea] giving itself the form of events" (Hegel, 2008: 317). The Idea is the absolute reconciliation of concepts and objectivity, it is the truth 'in itself and for itself'. The Idea is beyond the individual, but every individual is in some aspect associated with and part of the Idea.[3] For Hegel, the Idea (the Absolute) is the justification, the foundation of the desired organic connection between individual, society and state. The Hegelian 'Idea' has close similarities with the biblical 'Word' (in the Gospel according to St John), which is described as the essence and manifestation of divinity ("the Word was with God, and the Word was God"). Hegel understands reason and ethics as the infinite sources of the 'blessed water' needed for the advance of state and society together. The Hegelian political model can be, therefore, described as the quintessence of the project of European unification in the latter part of the twentieth century. A careful balance between flexibility and rigidity within the European Union institutions was needed in order to make the necessary shift from the Keynesian policies of the post-war time to the contemporary mixture of detailed regulation and liberalised socioeconomy. The successful handling of authority and liberties not only secured renewed economic prosperity (certainly disturbed by regular crises), but served as a magnet to attract new member states from southern and eastern Europe. At the core of the European Union system lie the

Hegelian lessons on politics, rights and the state. Hegel's ideas were highly instrumental in trying to resolve the old conundrum of the diversity between member state demands (from north to south, east to west) and the need to achieve higher levels of integration and common purpose. That could only be secured with a good deal of flexibility and the strengthening of a powerful European ideology. The European state is portrayed as the ultimate Idea, the source of reason and legitimacy, which is also sensible enough to develop novel areas of interaction and the application of law, in particular related to environmental matters. And, eventually, flexible environmental statehood is a direct beneficiary of Hegel's criticism of positive law and his defence of rational law as an improvement on tradition, violence and political rigidity.

Before the end of the twentieth century – the period of neoliberal globalisation when the state was tested to the limit – there were few opportunities to put the Hegelian state model into practice. It was the European Union project, particularly after the 1992 Maastricht Treaty, which offered the best prospects for the realisation (or actualisation) of Hegelian constitutionalism, especially in the realm of environmental regulation and management. In this process, environmental governance, instead of conventional government interventions, took centre stage in the pursuit of more flexible strategies and mechanisms of public administration that facilitate the accomplishment of socioecological goals, the realisation of values and the management of environmental risks and impacts. More recent negotiation between member states has redefined the EU as a large market arena embedded in a constitutional state framework that guards against competitive distortions, ensures efficient allocation of resources and prevents "undue government interventions in the market" (Kingston, 2011: 11). It is a form of liberalism, with some green tinges, that is cultivated and protected by the state, which continues to provide incentives, normalise behaviours and engage society in modernisation plans. In the EU, and in most of the world, environmental policies have become increasingly characterised by a more flexible association between market demands and environmental protection. For example, climate change policies shifted from a focus on charges and licences to the centrality of carbon trading and related pro-market schemes. Likewise, "corporate social responsibility" – defined as "a concept whereby companies integrate social and environmental concerns in their business operations and in their interaction with their stakeholders on a voluntary basis" (European Commission, 2002) – has turned into a new strategy for dealing with the negative public image of private companies because of environmental impacts. Particularly under the Fifth and Sixth Environmental Action programmes, the "use of economic instruments has gone hand-in-hand with emphasis on the integration principle and sustainable development at the EU policy-making level" (Kingston, 2011: 53), as demonstrated by the 2007 *Green Paper on Market-Based Instruments in Environmental Policy*. These instruments include fiscal incentives, tradable permits, and incentives for voluntary corporate initiatives, environmental management standards and eco-labelling.

Overall, the evolution to more flexible environmental statehood, as the combination of liberalism, clever regulatory instruments, higher levels of accountability and a renovated discourse, reveals the influence of Hegelian political theories underpinning EU environmental plans and programmes. Nonetheless, the ambitious plans of Hegel for the formation of a rational, ethical state have also been fundamentally constrained by the internal reasoning of his own philosophical argument, which eventually reinforces the failures of capitalist public policies and environmental regulation in particular. The Hegelian victory over absolutism and irrationality is, from the perspective of social and environmental justice, only a pyrrhic victory. That is because, following the Hegelian argument, the contemporary state is the historical necessity of a capitalist, market-based society, but it is also predicated upon those same relations and the myriad of socionatural impacts thus produced. As rightly pointed out by Negri (2011), the interior completeness of the Hegelian scheme has nourished and at the same time imprisoned the philosophical and political thought of the last two centuries. Moreover, Hegelian political elaboration is ultimately a 'conceptual trap' left in the pre-Victorian era for the distant future, the contemporary, post-Berlin Wall world. Hegel's constitutional schemes were certainly more advanced than other justifications of the bourgeois state, but only at the price of granting legitimation to a state system, and the associated economic order, based on the double exploitation of society and the rest of nature. The grand plan of a rational state sufficiently able to handle the contradictory forces of the capitalist socio-economy was as illusionary then as it is now.

Beyond the Hegelian constitutional plan

As discussed above, the present-day state, in charge of increasingly complex environmental problems, has operated through a careful combination of flexible regulation and the search for political legitimacy. The rearrangement of the state apparatus represented a unique opportunity for the implementation of the political theories advanced by Hegel almost two centuries earlier. The existing states, especially in the sphere of the European Union, constitute large administrative entities that constantly attempt to protect economic liberties in accordance with the imperatives of private property, commodity circulation and capital accumulation. In that sense, flexible statehood constitutes the culmination – making use of the Hegelian dialectics of flexibility and legitimacy – of the capitalist public administration's anti-commons agenda disguised as environmental conservation strategies. A critique of flexible environmental statehood requires a robust politico-ecological approach able to demystify the supposed superiority of flexible environmental statehood, as well as expressions such as sustainability, modernisation and governance. It is precisely here that a Marxian-informed analysis of environmental statehood can be of great help in identifying the shortcomings of Hegel's influence. Instead of an idealised political theorisation like the one advanced by Hegel, it is necessary to bring the politics of capitalist relations of production and reproduction to the centre of the examination of the state.

To begin with, there is nothing new in the controversies surrounding Hegel's state model. Despite the influence of Hegel's philosophy during his lifetime and in the decades following his death in 1831, acute criticism of the foundations of his political ideas had already emerged in the next generation of German thinkers. Feuerbach, for instance, criticised Hegel's 'absolute Idea' as nothing more than a deceased spirit of theology and a belief in pure phantoms (in Mehring, 1936: 52). Other authors in the same period blamed Hegel for his personal association with the Prussian State or for the mismatch between an innovative method and its reactionary conclusions. The most relevant criticism, especially because of his later work on the rules that govern capitalist relations of production – which incidentally helps to elucidate the socionatural contradictions of capitalism that permeate the state– was launched by Marx in his assessment of the Hegelian mystification of statehood and statecraft. Both Hegel and Marx believed in the superiority of the collective dimension over the individualism of natural law (*giusnaturalismo*), although for Hegel the higher political organisation of contemporary (capitalist) society has a positive meaning and for Marx the superstructure of capitalism has the opposite sense (Bobbio and Bovero, 1979). It is well known that, throughout his career, Marx made use of several Hegelian concepts, such as 'civil society' and 'property'; however, Marx cast them "in a revolutionary relationship to the concept of the state" (Avineri, 1968: 13). He identified internal contradictions in the Hegelian method, because of the dialectical 'kernel' that contrasts with the 'mystical shell' of his formulation (Colletti, in Marx, 1975). According to Marx, by inverting subject and predicate, Hegel tried to comprehend the state as an object which exists independently of the individuals that depend on the state. In particular, Marx censured the unqualified way in which Hegel accepted that the state, although imperfect or transitory, had direct correspondence with the perfection of the Idea. Marx gradually radicalised his political positions and rejected the Hegelian 'Idea' as the universal ontological category.

The continuities and disconnections between Marx and Hegel are complex and have been the object of interminable debate for more than a century, which need to be re-examined in the light of what has been learned about the state since the deaths of the two philosophers. For instance, Ilting (1984) believes that Marx overlooked that civil rights are the foundation of all duties of the Hegelian state and exaggerates the metaphysical tone of Hegel's work. Marx was apparently too concerned with the distortions created by the centrality of the Idea and failed to grasp that for Hegel the Idea is not only a form of thought, but both the expression of the concrete social reality and of the categories that correspond to, and help to create, that reality (MacGregor, 1984). Levine (2009) adds that Marx ignored Hegel's attribution of historicity to civil society, and mistakenly reduced the philosophy of Hegel to 'logical pantheism' and speculative idealism. In Levine's opinion, the way that Marx played down Hegel's sense of history and political practice would probably need to be recognised as an oversimplification of a more complex conceptual framework. To make matters worse, the main assessment formulated by Marx of the political theories of Hegel (i.e. Marx, 1970) is a highly fragmented text, especially because the manuscript was not intended for publication, but is in fact a notebook with notes and comments for personal use. This presents us with considerable analytical problems.

However, excesses aside, Marx was actually right in pointing out that Hegel formulated a highly idealistic model and had limited responses to questions of poverty, political exclusion and the centralisation of power in the hands of the emerging socioeconomic elite. According to Marx, Hegel committed so many antinomies and inconsistencies in the text of the *Philosophy of Right* that the end result was an argument that deceptively combines sophisms and speculation. Hegel refused to engage fully with the historical and political concreteness of the self-knowing and willing mind. Because of his largely idealist perspective, Hegel makes the objects independent merely by separating them "from their real independence, their subject" (Marx, 1970: 23). In contrast to Hegel, Marx rejected the view that the state could be described as an all-encompassing political community functioning according to a distinctive ethical character and acting as the fulfilment of reason in a world rife with inequalities and unreason. His view was quite the opposite; according to Marx, the capitalist state operates within the contradiction between multiple interests, but favours the stronger groups and classes. Marx (1975) later showed how Hegel reduced civil society to economic society through the mystified defence of the Christian state. Following Hegel's political philosophy, the perfect Christian state becomes the most perfect atheist state, to the extent that it is still theological but relegates religion to the level of civil society. Christian capitalism thus becomes the human basis of a flexible state that uses mystification to maintain the long-term anti-commons trends that underpin generalised socioecological exploitation.

Considering the main points of political mystification and the maintenance of socioeconomic hierarchies, the condemnation by Marx of the political philosophy of Hegel stood the test of time. Marx dedicated a full essay to Hegel's political system, in the form of a partially conserved manuscript written in 1843 and later published under the title *Critique of Hegel's 'Philosophy of Right'* (Marx, 1970). It is here that Marx demonstrates how the economic and political dimensions of the state are intrinsically, and necessarily, connected with class-based politics and ideological and political strategies to maintain value extraction and accumulation. Moreover, in order to fully understand the socioecological relevance of his more explicit political ideas, these need to be seen together with his other studies on the failure of conventional politics, economic inadequacies and society's interconnections with the rest of nature. Marx's early critical incursions, such as his doctoral thesis in 1841, constituted both an effort to go beyond the influential Hegelianism of the time (a task that was only partially fulfilled) and a return to the materialism of Epicurus (Foster, 2000). Marx saw that Epicurus had a non-conventional conceptualisation of the physical world (Mehring, 1936), particularly because of the acceptance of historical agency shared between nature and society (Giannotti, 2011). For Marx, a key problem with Hegel was the treatment of the 'essence of man' primarily located in the logical mind. Because Hegel wrongly perceived the world from a highly anthropocentric, Eurocentric and Lutheran perspective, any expression of the humanness of nature appears only as the product of a speculative mind. This inevitably leads to the estrangement, and the opposition, between abstract thinking and the sensuous reality (or the real sensuousness). Marx (1988: 148) condemned not "the fact that the human being *objectifies himself inhumanly*, in opposition to himself, but the fact that he *objectifies himself* in *distinction* from and in *opposition* to abstract thinking".

Building upon this estrangement between thought and reality, the Hegelian political model is fundamentally based on a dualism between civil society and the state. To a large extent, the previous Hobbesian tension between society and nature – i.e. positioning the latter as the realm of violence and brutality, and the first as the domain of peace and order – was replaced by the Hegelian separation between state and society. Although Hegel tried to bring together state and civil society, following the appeal for reason and liberty, he effectively created a fixed opposition and placed the state outside and above civil society. The primary loyalty of the Hegelian state is not with the communities of individuals, but with the fulfilment of core economic and political functions. Hegel developed an erudite conceptualisation of the state and of its metabolism with civil society (the economic realm) and with the family (the personal realm), but in the process he left what Plant (1973: 196) calls some "disruptive ambiguities which surround Hegel's philosophy of politics". For Hegel, the individual and the state are interconnected and interdependent, but this relationship can only happen if the rule of law preserves the existing institutions of private property and the stratification of political life (Hegel in *Realphilosophie*, quoted by Avineri, 1973). On the one hand, the state is proclaimed to be the manifestation of an ethical idea, the actualisation of freedom; but on the other hand, the supreme duty of the individual is to become a subordinate member of the state. The individual is dialectically sublated in the state, but only to have his or her needs contained within the preconceived boundaries of the 'rational' bourgeois state. This produces not simply a harmonic separation between private and public life, as claimed by Hegel, but an antagonism between the functions of the state (predicated as right in advance, as the state brings freedom and reason) and the actuality of private life (in Hegelian terms, logically containing a lower level of rationality).

Anticipating what later became reality (for instance, in the present-day politics of the European Union), Marx perceived great risks in the Hegelian state model and its tendency to serve mainly the interests of the propertied classes and their allied high bureaucracy. Despite its sophisticated logic, what Hegel envisages for the state means the perpetuation of the established political order and limited possibilities for real democracy. One of the core elements of Marx's critique of Hegel's dichotomic thinking is the idealisation of the role of the crown, seen as the centre of political legitimacy and the encapsulation of legitimate power over society. The Hegelian state is conceptualised as a complex organism that functions through estates, the executive and the crown, in which power is allegedly shared in a coordinated way that assures the perfect (rational) government. Hegel located the ultimate authority in the hands of the monarch, as the repository of maximum wisdom and independent judgment. Expressing his disapproval of Hegel's political plan, Marx claimed that "[t]he main thing is to fight against the constitutional monarchy as a hybrid creature, full of internal contradictions and bound to be self-destroying" (letter to Ruge on 20/3/1842, in Avineri, 1968: 9). Instead of democracy and general suffrage, Hegel wanted the individuals taking part in politics "as all", as a coherent group, rather than as individuals (Marx, 1970: 117). For Hegel, the resolution of social discrepancies should happen through the work of the Assembly of Estates (the parliament), which is the essential mediation between people and the political state. The legislature (i.e. the convergence of the estates) is thus seen as the totality of interests "not only in itself but also for itself" (Hegel, 2008: 287). Yet according to Marx, the sort of subjective freedoms announced by Hegel end up becoming formal, non-existent freedoms for the (majority) of the people.

The Hegelian political model is based on distorted dialectics where the individual – supposedly free and autonomous – is given the opportunity, and in practice required, to be incorporated (or sublated) in the Idea. In this way, the individual can retain his or her 'free will' and self-consciousness, but only at the cost of having to conform to a preordained plan of public affairs. Hegel proposes an idealised state – the perfect 'city' in the words of Plato – but insists on the importance of social and economic differences, as the catalysts of development and prosperity. Inequalities in terms of private property, as defended by Hegel, are not only necessary for the correct functioning of economic production, but serve as the embodiment of the opportunity for personal realisation. Socio-economic disparities are welcomed not only as an economic imperative, but as the necessary channel for the expression of universality. For Hegel, "the institutionalization of class relationships into the political structure is the way through which the atomism of civil society becomes integrated into a comprehensive totality" (Avineri, 1972: 104). Class differences – what Hegel described as primarily the estates of the peasantry, tradesmen and civil servants – correspond to disparities in the level of consciousness. Moreover, while Hegel recognised the centrality of private property and free markets, he saw the importance of reducing extreme hardship, promoting public education, a proper administration of justice and strict social control. If things go really wrong, Hegel's recommendation is ultimately geographical: expand colonisation and transfer part of the population to other parts of the world (possibly where new resources can be appropriated and incorporated into capitalist relations of production and inevitably reproduce the socioecological problems of the motherland). The urban poor (the 'rabble') is formed by the marginal individuals who form a class and are believed to have no chance of ever becoming an estate with political representation. Their pre-given political ineptitude, therefore, becomes the perpetual justification for their social and economic exclusion.

The Hegelian claim that the state is rational has the perverse result of allowing the expansion of an economic system that, in practice, allows only minimal levels of individual freedom (including the freedom to protest against environmental degradation). In that sense, Hegel actually advocated a neo-Platonic state that is adjusted to the demands of a mass production and mass consumption society. For Hegel, just as for Plato, plain democracy leads to lawlessness, anarchy and departure from the ideal state, eventually preparing the ground for the emergence of tyrannies. To avoid this, the state envisaged by Hegel is expected to have a double impact on the individual: it is something external to him and it is the promise of the realisation of his or her own rational ends in life. The individual is both constrained by the state and simultaneously supposed to be moving towards freedom (or at least believes that his or her freedom is increasing). In the words of Reyburn (1921: 235), the "restraints of public life are the articulations which the state requires in order to attain its proper unity and organization, and the citizen who is conscious of his identity with the state is made free by them." Hegel (in close association with Kant) put aside the need for plain democracy and social equality, because it is not the individual that needs to be represented in the state but his or her interests that can be presumed and captured through rational thinking. In the *Encyclopaedia*, Hegel argued that 'mind' has stepped out of mere subjectivity, but "the full realization of that freedom, which in property is still incomplete and formal, is reached only in the state, in which mind develops its freedom into a world posited by it, an ethical world", as much as mind must also help again to free the world (Reyburn, 1921: 265).

The Hegelian compromise between essence and existence – that is, the conciliation between the rational ontology of the state (the manifestation of reason) and the ethical practice expected from people and governments (a gradual, historical advance towards freedom and development) – produced one of the most enduring, but deceptive, reasonings associated with statehood. This compromise was rejected by Marx due to the inverted logic of its proposition: instead "of having subjects objectifying themselves in public affairs Hegel has public affairs becoming the subjects" (Marx, 1970: 62). These distortions were experienced again when environmental statehood was expanded and reorganised in the latter decades of the twentieth century. The establishment of a dedicated nucleus of environmental policy and regulation represented one of those contradictory attempts to renovate the state apparatus in the search for legitimacy and flexibility. Instead of merely anticipating the autocratic states commanded by Bismarck and Hitler (as mistakenly claimed by Popper, Russell and others), the most important consequence of Hegel's political model was to offer the intellectual tools for the conservative and superficial renovation of the state. Informed by Hegelian constitutionalism, the environmental initiatives promoted by the state never really challenged the underlying appropriation of resources and ecosystems because of the politico-economic commitments of the state apparatus and its primary association with the hegemonic groups of interest. As rightly observed by Lefebvre (2009: 84), the Hegelian state "does not arbitrate conflicts, it moderates them by keeping them within the limits of the established order."

For all the above reasons, a critique of the Hegelian constitutional plan has major repercussions for the politico-ecological examination of environmental statehood, as this book has shown. The Hegelian defence of economic inequalities and the emerging capitalist state has had a profound impact on the development of environmental statehood. The unsustainability of the formal sustainability agenda (i.e. the contradiction between the environmental discourse and the socio-economic practice of the state and its main allies) can be explained by the Hegelian relationship between the real and the rational (i.e. the real is rational by definition). Following Hegel's plans for the state, claims for 'ecological protection' have been advanced by politico-economic structures responsible for systematic and widespread impacts. The environmental regulation of the capitalist state has its foundation in what Hegel called the 'police' [*Polizei*] or the affirmation of public authority required to contain conflicts and social distress, although it is a form of authority that is supposed to emanate from the legitimacy of the power structures (Hegel, 2008).[4] The political regime bequeathed by Hegel has been instrumental in limiting democracy and social justice within the boundaries of private property, unfair production relations and stratified governance. Equally, the reduction of social and political differences to the 'common' language of money (a key tenet of present-day ecological modernisation) was already suggested by Hegel as a mechanism that could strategically forge collaboration (see Hegel, 2008: 285), while in effect it serves to exacerbate socioecological tensions.

According to Marx, the capitalist state that follows the Hegelian political model turns out to be the perpetuator of a domination of exchange value over use value. Following Hegel, the capitalist state is not only a class instrument, but also the cornerstone of the maintenance, and constant updating, of the economic order of capitalist society. This order is essentially based on the anti-commons priorities of specific relations of property and production. For Hegel, the only legitimate form of property recognised by modern society is private property and the ideal social formation resembles the bourgeois preferences (Wood, 1993). "The State as capital [with the main purpose of controlling social labour] is the State as development. Power is exercised in development, by development" (Negri, 2011: 38). Hegel superimposed an ideal (and idealised) constitution of the state on a world that was becoming increasingly capitalist, where collective problems derived from economic acceleration were challenging the old, authoritarian state. As a result, Hegel responded to the need to preserve rationality in an uncertain economic reality fraught with the dissatisfaction of the majority of the population. This is particularly evident in the *Philosophy of Right* where Hegel formulates "the supreme index of bourgeois ideology and the capitalist practice of the organization of exploitation" (Negri, 2011: 44). It is here that the Hegelian state is best described as a complex entity, a rational organism that operates and creates its own moral and political contingencies through the careful, selective inclusion of citizens as members of the state (Hegel, 2008). As such, the individuals are entitled to participate in political life, but only within the narrow sphere of preordained assemblies and under the command of the monarch, the ultimate guarantor of harmony between all contingencies. Hegel advocated the dialectical abolition of the existing configurations of state and civil society, but his project condemns the replacement of the public arena with rationalised state procedures (as representation of the ethical whole). Hegel based his argument on ethics and morality, but was very conservative about social change, which could only happen through education and under the leadership and reflection of the 'philosophers' (Wood, 1993).

The Hegelian constitutional ideas on legitimacy, social inclusion and political stability have certainly permeated the renovation of the European state system under the pressures of globalisation and neoliberalism. In particular, they provided the theoretical and moral justification for the consolidation of flexible environmental statehood. This means that the Hegelian ontology of the modern state allowed, and even presupposed, the movement from the conventional to the flexible configuration of environmental statehood and regulation. For Hegel, the state is an integral part of the ethical life of a people and it should occupy the middle ground between authoritarianism and anarchy. It should be the expression of reason (which is a concrete concept, directly related dialectically to actual political behaviour), flexibility and harmony within established sociopolitical structures. In that sense, Hegel anticipated the contemporary argument in favour of environmental governance, gradual adjustments and controlled public participation. For example, the OECD (2011) justifies the importance of new ways of dealing with water resources and ecosystems as responding to the need to secure equal and universal public water services for all. The modernisation of water management thus requires operational flexibility through the (dialectical) increase of regulation. In addition, the organisation emphasises the importance of addressing local geographical differences within multi-level governance: "every country must find its own balance among the three basic sources of finance in the water sector"; that is, the combination of service tariffs, international transfers and government investments and subsidies (OECD, 2011: 63). These three sources are the Holy Grail of water governance, the proclaimed '3 Ts' of taxes, transfers and tariffs. This perspective on water governance candidly reveals the Hegelian recommendation in terms of the high moral ground of state interventions and an increased rationality (closely associated with economic thinking) that connects the various branches of the state apparatus. Making use of Hegel's political framework, the governance agenda intends a legitimate and unified state action, but always from the perspective of those in control of the state.

Bustamante *et al.* (2012) provide another demonstration of the lasting, subtle legacy of the Hegelian trap and its expansion to socionature in the non-Western world. In Bolivia, the left-wing government of Evo Morales has formally institutionalised the right to water in the new national constitutional order under the claim that it was part of its pro-Indian and pro-poor policies. Although on paper it seems an important democratic measure, Bustamante *et al.* (2012) strongly criticise the simplistic discourse on rights and their manipulation as part of the intensified exploitation of water resources (aiming, in particular, to sell hydroelectricity to Brazil and to safeguard the interests of the larger irrigators) and the promotion of industrialisation (the so-called Great Industrial Leap). In this case, a democratically elected government, with unique rhetorical commitments to the Bolivian poor, has ended up operating within a spurious Hegelian logic and, as a result, imposing mediation between state, nature and society that prioritises the conventional model of economic growth and private property accumulation. The Morales government cannot be blamed for policies that are simply utilitarian and pro-capitalism (vis-à-vis his confrontation of the international capitalist order and concrete poverty alleviation measures), but can certainly be criticised for its Hegelian mystification of the balance of rights and gains. From this example it can be inferred that Hegel provided enough flexibility for the national state to adapt and mutate to other geographical contexts (way beyond the European and German circumstances of the first half of the nineteenth century).

A comparable example comes from the discussion surrounding the revitalisation of nuclear electricity in the UK that is taking place at the time of writing (the years 2012 and 2013), considered by many politicians and academics to be 'a necessary evil' in order to maintain the established standards of life and commodity consumption. The argument in favour of new nuclear power plants is more than mere utilitarianism (although the frequent comparison between the lesser evils of nuclear energy and the apocalyptic consequences of climate change betray a clear utilitarian rationale): the advocates of British nuclear energy invoke a Hegelian reasoning that the technology is 'reasonably' safe and, in any case, radioactive rubbish would only become a problem in the very distant future. Following this line of argument and considering that human society (at least in a more organised form) is no more than 10,000 years old, how could we possibly accommodate moral claims that will be important thousands of years ahead of us? The Hegelian political model is again organised around the state in charge of the capitalist world and concerned with the short-term impact of environmental degradation. The capitalist state continues to be defended as the supreme incarnation of reason and, despite the daily criticism of politicians and state bureaucrats, its foundations are not really open for discussion. Following the Hegelian logic, to "wish to alter this State or subvert it was morally bad, because directed against the rational will embodied in it, and in any case futile, because set against a decision made by history" (Berlin, 1963: 64).

Therefore, it is not possible to agree with Ilting (1984) that Hegelian constitutionalism is now useless as a political programme. On the contrary, Hegel's state model has had long-lasting consequences and has had a particular influence on the implementation of flexible environmental regulation approaches. Hegel's attempt to reconcile flexibility and firm authority paved the way towards accommodating, within the same rigid and hierarchical dialectics, the reorganisation of flexible environmental regulation and policy-making. The mystification of the state, introduced by Hegel during a period intensely disrupted by the Napoleonic Wars and the emergence of the bourgeoisie as the decisive political force, arises again in the context of environmental governance and sustainable development. The highly technified treatment of environmental issues in the early twenty-first century actually betrays a profound Hegelianism, in the sense that the state is expected to make sense of challenging socionatural issues, but only through the maintenance of anti-commons, and ultimately anti-ecological, priorities. The contemporary state, informed by the Hegelian theories, adopted a 'greener' appearance, but lost the ability to undertake changes in core political commitments. Any calls for pluralism and democracy that avoid tackling the Hegelian idealism of the current day environmental responses (e.g. Eckersley, 2004) tend to reinforce and, consequently, perpetuate the socioecological contradictions of flexible environmental statehood. The Hegelian construction of individual consciousness rests on the worship of the nation-state and on the cult of private life and mass consumption (Greisman and Mayes, 1977), but the solutions to collective problems advanced by Hegel are inherently ambiguous and instrumental in the hands of state authorities.

The contradictions of flexible environmental statehood, such as the environmental agenda adopted by the European Union since the 1980s, can be better appreciated with the help of Marx's reading of the Hegelian idealisation of the state: it is the paradox of achieving all and nothing at the same time. The environmental action of the contemporary state is in effect shrouded in mystification, elitism and manipulation of public affairs for the benefit of those in control of the state. The application of detailed science, parliamentary lawmaking and systematic public consultation may give the impression that the state is effectively moving towards higher levels of sustainability and ecological citizenship. However, these are highly contained and carefully rationalised approaches to environmental problems that aim to contain the declining profitability of some economic sectors and create novel opportunities for the accumulation of capital. Furthermore, despite the ingeniousness of the Hegelian political model, the responsive capacity of flexible environmental statehood is also increasingly showing signs of inadequacy and exhaustion. Hegel's philosophical system in fact negates itself, because, contrary to his own understanding, it is "based upon the perennial resistance of the nonidentical" (Adorno, 1973: 120). The identity, as totality, takes ontological precedence, but it is an artificial identity that is promoted as the absolute concept. Theory serves to force, to impose indissolubility, rather than bringing the indissoluble as the outcome of meaningful reasoning. In the end, Hegel's dialectic becomes suspicious of all identity, its "logic is one of disintegration" (Adorno, 1973: 145).

Marx (1975), in the article *On the Jewish Question*, offered a compelling alternative to the Hegelian mystification of the state and its containment by hegemonic politico-economic interests. In his analysis of the prospects of religious freedom, Marx subverted the conventional political argument of the time, which advocated political emancipation from a highly conservative perspective. In Marx's opinion, the political emancipation of the religious person requires, first of all, the emancipation of the state from religion. In other words, it was not enough to secure additional political rights "*within* the prevailing scheme of things", especially when contained by private property relations and the modern state (Marx, 1975: 221). On the contrary, in order to become a genuine social being (instead of only an abstract citizen), the individual needs to secure his or her human emancipation, which means overcoming the separation of the individual from social forces. There is a fundamental message here, which can be used to explain the fundamental environmental contradictions of the neoliberal state. For Marx, the perfect Christian state is the atheist state (such as the United States), which granted religious freedom as the artefact to promote and consolidate a highly religious society. In the case at hand, flexible environmental statehood, with its green discourse and complex regulatory apparatus, has become the champion of widespread environmental degradation and socionatural impacts. Marx identified this early contradiction in the political agenda of the state and, eventually, convincingly disarmed the Hegelian argument.

The fundamental antinomies of flexibility and legitimacy of contemporary environmental policies need to be situated in this wider politico-ecological debate about the substantial transformation of the current state apparatus. The key onto-logical and political question is less how the state deals with the environmental policies and strategies per se, but what its ultimate commitments are and how it works to reinforce or eliminate processes of exclusion and exploitation. As observed by Marx (1975), it is the manner of emancipation that needs to be criti-cised, given that the state can liberate itself without people being set free. The fundamental distortions of conservative democracy "must be looked for in the *nature* of the state itself" (Marx, 1975: 217). This requires sustained and radical changes in small, specific state practices and also in wider commitments and interventions of the state. What is more, the renewal of existing state formations should happen both from the outside and from the inside of the state. The tran-scendence of the dualism between structure and agency of the state requires the avoidance of nature–society dualisms, which are to a large extent promoted and reinforced by the state itself. In the end, those multiple answers to socionatural disputes around the state should become a main unifying catalyst that brings together sociopolitical emancipation and a just, socioecologically viable, society.

Notes

1 The notion of an 'ethical community' by Hegel evidently echoes Aristotle's claim of the Greek *polis* as an integrated community ("There is one thing clear about the best constitution: it must be a political organization which will enable anyone to be at his best and live happily", Aristotle, 1995: 255).
2 Interestingly, in a remarkable (perhaps even proto-ecological) critique of the Württem-berg Estates, Hegel concluded that their fundamental error was to start from positive law, acting "like a landed proprietor whose sandy soil has been covered by fertile humus as a result of a beneficent flood and who yet proposed to plough and farm it exactly as he had done before" (Hegel, 1964: 281).
3 We can compare Hegel's attempt to produce an all-embracing philosophy that recon-ciles, in a humanist fashion, the spiritual world with the physical reality, with Father Caspar trying to explain where God found all the water needed for the Great Flood (in Umberto Eco's 'The Island of the Day Before'). After a long biblical and geographical examination, Caspar concludes "what no other human being before now had thought ever", that water could only be obtained at the point where today and tomorrow coin-cide. At this point, there is an apparently infinite source of water, given that yesterday's water can be used today, and today's water can then be poured tomorrow. In order to prove his fabulous discovery, Caspar instructs Roberto to adopt a celestial reasoning: "So then you try to think what you do if you are God" (Eco, 1998: 262–3). This is Caspar's method, as much as Hegel's.
4 Rancière (2007) expands this concept of the 'police' to express what is normally described as politics. In his view, the 'police order' is the process of governing that prescribes, instead of protecting, a given reality or social sensibility (regarding the underlying norms that define what is allowed or not allowed, available or unavailable in a given situation).

Bibliography

Adorno, T.W. 1973. *Negative Dialectics*. Trans. E.B. Ashton. Routledge and Kegan Paul: London.

Aristotle. 1995. *The Politics*. Trans. E. Barker. Oxford University Press: Oxford and New York.

Avineri, S. 1968. *The Social and Political Thought of Karl Marx*. Cambridge University Press: Cambridge.

Avineri, S. 1972. *Hegel's Theory of the Modern State*. Cambridge University Press: Cambridge.

Avineri, S. 1973. Labour, Alienation, and Social Classes in Hegel's *Realphilosophie*. In: *The Legacy of Hegel*, O'Malley, J.J., Algozin, K.W., Kainz, H.P. and Rice, L.C. (eds). Martinus Nijhoff: The Hague, pp. 196–215.

Berlin, I. 1963. *Karl Marx: His Life and Environment*. 3rd edition. Oxford University Press: London.

Bobbio, N. 1995. *Stato, Governo, Società: Frammenti di un Dizionario Politico*. Einaudi: Torino.

Bobbio, N. and Bovero, M. 1979. *Società e Stato nella Filosofia Politica Moderna: Modello Giusnaturalistico e Modello Hegelo-Marxiano*. Il Saggiatore: Milano.

Bustamante, R., Crespo, C. and Walnycki, A.M. 2012. Seeing through the Concept of Water as a Human Right in Bolivia. In: *The Right to Water: Politics, Governance and Social Struggles*, Sultana, F. and Loftus, A. (eds). Earthscan: London and New York, pp. 223–40.

Dickey, L. and Nisbet, H.B. (eds). 1999. *Hegel: Political Writings*. Trans. H.B. Nisbet. Cambridge University Press: Cambridge.

Eckersley, R. 2004. *The Green State: Rethinking Democracy and Sovereignty*. MIT Press: Cambridge, MA and London.

Eco, U. 1998. *The Island of the Day Before*. Trans. W. Weaver. Vintage: London.

European Commission. 2002. *Commission Communication of July 2, 2002 Concerning Corporate Social Responsibility: A Business Contribution to Sustainable Development*. COM (2002) 347 Final.

Foster, J.B. 2000. *Marx's Ecology: Materialism and Nature*. Monthly Review Press: New York.

Giannotti, J.A. 2011. *Marx Além do Marxism*. L&PM: Porto Alegre.

Gidwani, V. 2008. The Subaltern Moment in Hegel's Dialectic. *Environment and Planning A*, 40, 2578–87.

Greisman, H.C. and Mayes, S.S. 1977. The Social Construction of Unreality: The Real American Dilemma. *Dialectical Anthropology*, 2(1–4), 57–67.

Harvey, D. 2005. *A Brief History of Neoliberalism*. Oxford University Press: Oxford.

Hegel, G.W.F. 1953 [1824]. *Reason in History, a General Introduction to the Philosophy of History*. Trans. R.S. Hartman. Liberal Arts Press: New York.

Hegel, G.W.F. 1964. *Hegel's Political Writings*. Intro. Z.A. Pelczynski; trans. T.M. Knox. Oxford University Press: Oxford.

Hegel, G.W.F. 1969 [1812]. *Science of Logic*. Trans. A.V. Miller. Humanities Press International: Atlantic Highlands, NJ.

Hegel, G.W.F. 1977 [1807]. *Phenomenology of Spirit*. Trans. A.V. Miller. Oxford University Press: Oxford.

Hegel, G.W.F. 2008 [1821]. *Outlines of the Philosophy of Right*. Trans. T.M. Knox. Oxford University Press: Oxford.

Humboldt, W. 1993 [1791–2]. *The Limits of State Action.* Trans. J.W. Burrow. Liberty Fund: Indianapolis, IN.

Ilting, K.-H. 1984. Hegel's Concept of the State and Marx's Early Critique. In: *The State and Civil Society: Studies in Hegel's Political Philosophy*, Pelczynski, Z.A. (ed.). Cambridge University Press: Cambridge, pp. 93–113.

Kingston, S. 2011. *Greening EU Competition Law and Policy.* Cambridge University Press: Cambridge.

Knowles, D. 2002. *Hegel and the Philosophy of Right.* Routledge: London.

Lee, D. 2008. The Legacy of Medieval Constitutionalism in the *Philosophy of Right*: Hegel and the Prussian Reform Movement. *History of Political Thought*, 39(4), 601–34.

Lefebvre, H. 2009. *State, Space, World: Selected Essays.* Trans. Moore, G., Brenner, N. and Elden, S. University of Minnesota Press: Minneapolis, MN.

Levin, M. and Williams, H. 1987. Inherited Power and Popular Representation: a Tension in Hegel's Political Theory. *Political Studies*, 35, 105–15.

Levine, N. 2009. Hegelian Continuities in Marx. *Critique: Journal of Socialist Theory*, 37(3), 345–70.

MacGregor, D. 1984. *The Communist Ideal in Hegel and Marx.* University of Toronto Press: Toronto.

Mann, G. 2008. A Negative Geography of Necessity. *Antipode*, 40(5), 921–34.

Marx, K. 1970 [1843]. *Critique of Hegel's 'Philosophy of Right'.* Trans. A. Jolin and J. O'Malley. Cambridge University Press: Cambridge.

Marx, K. 1975. *Early Writings.* Trans. R. Livingstone and G. Benton. Penguin and New Left Review: London.

Marx, K. 1988 [1930]. *Economic and Philosophic Manuscripts of 1844.* Trans. Martin Milligan. Prometheus Books: Amherst, NY.

Mehring, F. 1936. *Karl Marx: The Story of His Life.* Trans. E. Fitzgerald. John Lane: London.

Nancy, J.-L. 1997. *Hegel: L'Inquiétude du Négatif.* Hachette: Paris.

Negri, A. 2011. Rereading Hegel. In: *Hegel and the Infinite: Religion, Politics, and Dialectic*, Žižek, S., Crockett, C. and Davis, C. (eds). Columbia University Press: New York, pp. 31–46.

OECD. 2011. *Water Governance in OECD Countries: A Multi-Level Approach.* OECD Studies on Water. OECD Publishing.

O'Hagan, T. 1987. On Hegel's Critique of Kant's Moral and Political Philosophy. In: *Hegel's Critique of Kant*, Priest, S. (ed.). Oxford University Press: Oxford, pp. 135–59.

Pelczynski, Z.A. (ed.). 1984. *The State and Civil Society: Studies in Hegel's Political Philosophy.* Cambridge University Press: Cambridge.

Plant, P. 1973. *Hegel.* George Allen & Unwin: London.

Popper, K.R. 2002. *The Open Society and its Enemies.* Routledge: London.

Rancière, J. 2007. *On the Shores of Politics.* Trans. L. Heron. Verso: London and New York.

Reyburn, H.A. 1921. *The Ethical Theory of Hegel: A Study of the Philosophy of Right.* Clarendon Press: Oxford.

Rockmore, T. 1989. Modernity and Reason: Habermas and Hegel. *Man and World*, 22, 233–46.

Russell, B. 1950. *Unpopular Essays.* Allen & Unwin: London.

Sager, F. and Rosser, C. 2009. Weber, Wilson, and Hegel: Theories of Modern Bureaucracy. *Public Administration Review*, 69(6), 1136–47.

Smith, S.B. 1989. *Hegel's Critique of Liberalism: Rights in the Context.* University of Chicago Press: Chicago and London.

Stewart, J. (ed.). 1996. *The Hegel Myths and Legends.* North-Western University Press: Evanston, IL.

Taylor, C. 1979. *Hegel and Modern Society.* Cambridge University Press: Cambridge.

Toews, J. 1993. Transformations of Hegelianism, 1805–1846. In: *The Cambridge Companion to Hegel,* Beiser, F.C. (ed.). Cambridge University Press: Cambridge, pp. 378–413.

West, D. 2012. Continental Philosophy. In: *A Companion to Contemporary Political Philosophy,* Goodin, R.E., Pettit, P. and Pogge, T. (eds). Wiley-Blackwell: Oxford, pp. 36–68.

Wood, A.W. 1993. Hegel and Marxism. In: *The Cambridge Companion to Hegel,* Beiser, F.C. (ed.). Cambridge University Press: Cambridge, pp. 414–44.

Žižek, S. 2003. *The Puppet and the Dwarf: The Perverse Core of Christianity.* MIT Press: Cambridge, MA.

Žižek, S. 2006. *The Parallax View.* MIT Press: Cambridge, MA.

Žižek, S. 2011. Hegel's Century. In: *Hegel and the Infinite: Religion, Politics, and Dialectic,* Žižek, S., Crockett, C. and Davis, C. (eds). Columbia University Press: New York, pp. ix–xi.

7 What is beyond flexible environmental statehood and the naïve faith in the eco-state?

This final chapter summarises the argument of the book and identifies the main lessons learned about the political ecology of the state. After the long journey through conceptual and empirical issues, it is hopefully easier now to understand the aims and constraints of environmental statehood that have characterised the history of the mainstream responses to environmental problems during the last hundred years or so. It should also be evident that, despite the number of policies, agencies and procedures dedicated to dealing with environmental problems, there is very limited indication that the trend of environmental disruption has been lessened. On the contrary, old and new environmental questions continue to undermine social and economic activities, as well as affecting the long-term viability of ecosystems on the planet. The main reason for this persistent failure is that the state has attempted to mitigate and resolve many negative impacts, but it has followed the asymmetry of political power and its own hegemonic commitments while doing so. Consequently, the state has turned out to be the main environmental player of the contemporary world, but its structure and operation have become the main locus of socioecological conflicts and permanent contestation. In addition, its various environmental initiatives have been largely permeated by an exogenous rationality based on the separation between society and the rest of socionature, as well as on economic and anthropocentric ideologies. A politico-ecological approach is, therefore, necessary to permit a proper comprehension of the intersectoral disputes and multiscale barriers responsible for the systematic reproduction and continuation of environmental problems.

The contradictory basis of environmental statehood ultimately derives from the anti-commons character of the capitalist state, which was justified early on by Adam Smith, and then reinforced by generations of both liberal and welfare economists. The private property of production and the accumulation of wealth were considered essential elements for the good functioning of a capitalist economy. In that context, the state was expected to maintain safeguards against the enemies of privatisation and the defenders of the commons. What was not evident for those economists was that the expansion of urban–industrial activities, as well as the intensification of mining and agriculture production, would inevitably lead to serious levels of socioecological degradation. Especially since the post-Second World War period, there has been considerable dissatisfaction with the rate of socioecological impacts, and mounting reactions from communities and social groups more directly affected. The welfare, developmentalist state had to mediate in these environmental conflicts, but at the same time had to maintain the supply of resources and energy, preserve the profitability of economic activities and encourage novel mechanisms of production and capital accumulation. The range of material and symbolic initiatives employed during the twentieth century by the state to cope with socioecological problems became a new area of public policy-making and part of the overall process of statecraft (which went from liberalism to Keynesianism and then into the neoliberal reform of the state).

This book initially examined the origins of environmental statehood and its subordination to wider socioeconomic demands and Western patterns of reasoning. Environmental statehood, as an array of institutional arrangements and ideological approaches, became an integral component of the processes of economic production, as the response to political tensions and the bridge between capitalist and non-capitalist forms of organisation. While the economy evidently relies on and benefits from the richness of socionature – which is incidentally transformed into resources and services by the expansion of capitalist activities – the economic agents acting alone typically over-exploit many elements of socionature beyond the point of ecosystem recovery. This form of rational irrationality is a distinctive feature of the capitalist economy. Under a non-capitalist regime with mainly shared resources and more explicit socionatural interdependencies, those limits are imposed and enforced by the whole community taking into account the long-term viability of the economic system.[1] However, capitalist relations of production reduce or abolish common property institutions, leaving a vacuum that needs to be filled by the state (in its role as the regulator of private property interests). Moreover, the environmental role of the state is not simply to provide mediation and control for the greater good of the economy. The state actually assumes the role of a biased referee that is not capable, and has no intention, of completely removing itself from socionatural processes and associated disputes. Instead of detached or purely technocratic action, the capitalist state operates through its inescapable interdependencies with society and the rest of socionature. This is because the reality of the world is profoundly trialectical – society, state and socionature as a single, dynamic entity – and it unfolds from the micro to the macro scales of interaction.

Following this trialectical ontology, the state is simultaneously at a distance from and also unavoidably connected to society and the wider socionature. Crucially, this 'impure' role of the state is instrumental in both supporting the expansion of capitalist production and promoting the recovery of degraded ecosystems and resources. In other words, the environmental responses of the state encapsulate both its anti-commons and pro-commons responsibilities according to the specific historic–geographical circumstances and the balance of political power at the time. Since the turn of the twentieth century, successive models of environmental statehood were adopted by Western countries and then replicated around the globe. The ideological and political elements of environmental statehood were translated into more visible and tangible regulatory instruments, such as dedicated agencies, policies and legislation, which constitute specific state-fixes. Environmental statehood has been implemented as state-fix because the reactions to environmental problems have always been contingent and provisional. The reason for the resort to temporary state-fixes is that the contemporary (capitalist) state is actually unable to produce effective and lasting solutions to socioecological problems that have their origin in the very economic system it is primarily meant to protect. Multiple state interventions always assume the format of transient state-fixes set up for dealing with the most serious or pressing problems, but only on condition that these state-fixes do not impair the overall pattern of mass consumption, mass wastage and socionatural exploitation.

The first model of environmental statehood coincided with the macro-economic and technological adjustments taking place during the Second Industrial Revolution. This introductory mechanism of environmental statehood was, however, too patchy and superficial to secure any significant changes in the processes of production and consumption. This initial experience was followed by the conventional model of environmental statehood, synergistically connected with Keynesian development policies. It was accordingly characterised by larger administrative structures and detailed systems of regulation and control. But the initiatives associated with conventional statehood were soon considered by most private sector companies to be too bureaucratised and ineffective. Conventional statehood was condemned for being too onerous to private businesses and for failing to contain accelerating trends of socioecological degradation. This model of environmental statehood corresponded to what (Luke, 1990: 159) describes as the "mono-dimensional interventions" of the welfare state in its struggle to recover production and accumulation. A few decades later, at the turn of the twenty-first century, a more flexible model of environmental statehood was introduced as an effort to optimise environmental regulation and engage larger sectors of society. Flexible environmental statehood had a contingent relation with the overall renewal of the state apparatus that followed neoliberal pressures. It also facilitated the growing neoliberalisation of socionature through a series of procedures intended to generate profit directly from environmental management (e.g. certification, privatisation, commodification, commercialisation, etc.) rather than simply from the traditional exploitation of natural resources.

As discussed in Chapters 3 to 6, the trajectory of environmental statehood reveals the distinct influence of some of the most important modern theorists who had already informed the evolution of wider political and constitutional affairs in Europe and beyond. The ideas of Hobbes, Kant and Hegel had an especially decisive imprint on the launch and functioning of environmental statehood. Hobbes's definition of nature as the realm of brutality and irrationality, together with his advocacy of a strong, centralised authority, provided the basis of the early and conventional environmental statehood based on command-and-control strategies. The transition to less interventionist and rationalised approaches (still within the sphere of private property and liberal economy) was largely inspired by the ideas of Kant. But the pinnacle in the evolution of

environmental statehood was the search for governance, sustainability and ecological modernisation, which was supported by well-crafted discourse on participation, harmony and legitimacy. This last, and still prevailing model of
environmental statehood bears the hallmark of Hegel's political theories and his
advocacy of a harmonious coordination between the different branches of the
state and the representatives of the most influential social groups. It has also
been argued throughout this book that, in order to understand the shortcomings
of the three mainstream models of environmental statehood, a political ecology
perspective can greatly benefit from Marx's critique of the economic and socioecological contradictions of the capitalist society, as well as from his attack on
the idealisation of the state.

The present-day organisation of environmental statehood is predominantly
informed by the Hegelian theses on legitimacy, integration and hierarchy. The
philosophy of Hegel provided, even indirectly, the intellectual tools needed to
update the outmoded regulatory approaches and prolong the life of flexible
environmental statehood. In contrast to Hobbes's request for a powerful authority to contain the 'state of war' and the related miserable condition of every man
against every man, the Hegelian state is an entity with high moral standards and
legitimised actions. The ambiguous, but highly persuasive, argument of Hegel
allowed the superimposition of a new model of environmental statehood since
the end of the 1980s – portrayed as the expression of administrative wisdom,
public engagement and scientific aptness – upon a socionatural reality fraught
with lasting environmental impacts and associated conflicts. However, even if
Hegelian political ideas may have served to improve and better justify the flexibilisation of environmental statehood, the contradictions and inadequacies nonetheless persist. The reconciliation of the individual with the state, as well as with
the community and the environment, was a task that Hegel set up in his philosophy of politics (Plant, 1973). Hegel was particularly concerned with a rational
defence of the capitalist state, but in a way that also allowed some degree of subaltern freedom to the individual and entrepreneurialism. For him, corporations
(i.e. organisations of particular industries) constitute a nucleus of legitimacy and
political activity that help to bind together the state, society and the individual.
Hegel detected a major problem in the ordinary individualism of bourgeois
society and, therefore, strived to mitigate it with elements of the pre-bourgeois
order, such as the remaining institutions of the medieval guilds.

On the whole, the nuanced political thinking of Hegel permitted the subtle,
but deliberate, suppression of political liberties so that society and the economy
could be modernised according to the tenets of the capitalist ideology of production and consumption. Hegel cautioned against a strict laissez-faire economy, as
well as against too much state interventionism. His political dream seems to
envisage a population made up of a sort of semi-autonomous individuals, who
voluntarily accept the authority of the state and are amenable to the consolidation of conservative adjustments. "Hegel's definition of civil society clearly
follows the model of the free market in which it is 'every man for himself' and a
Smithian 'invisible hand' ensures that all will turn out for the best" (Cullen,

1979). This was something rather innovative in his own time and, more importantly, anticipated some of today's debate on public participation, freedom of information and flagship policy initiatives. Such ingenious, but highly contradictory, logic pervades Hegel's entire political elaboration, which is evident in *Outlines of the Philosophy of Right* and is even more unmistakably present in his *Realphilosophie* (a publication that contains lectures given at the University of Jena, between 1803 and 1806, which comprise detailed discussions of poverty, labour exploitation and wealth disparities).

In *Realphilosophie*, the state is defined "as the transcendence of the individual into the universality of the law"; because the individual is simultaneously a particular human being and a universal person (i.e. both a member of society and a citizen of the state), the action of the state is both 'instrumental' and 'immanent' (Avineri, 1973: 209). Hegel's ultimate aim was to bring together, with a minimal level of conflict and maximised legitimacy, subjective freedom, private property and state rationality for the bourgeois development of Prussian society. The Hegelian state model is a concerted attempt to reintegrate the self into the universal being, especially in moments of disruption and chaos caused by economic activity (Avineri, 1973). However, although it may have been a relatively advanced proposition for a society still dominated by a decadent aristocracy, the Hegelian solution was nonetheless primarily meant to boost the leadership and authority of the new social elite, who could combine multiple alliances between the emerging bourgeoisie and the old aristocratic rulers. Hegel was thinking from the perspective of those that wanted Prussia to emulate Britain and France in terms of their industrial and imperialist expansion. That is why Marx saw in Hegel the theorist of the modern representative state, someone who articulated the aspirations of the Prussian State in its attempt to escape backwardness.

For Marx (1970), the crux of the matter is that the Hegelian state model can only operate through the 'mystification of reason', that is, the metaphysical identity between the real and the rational. One unfortunate outcome of such a proposition is that, when confronted with a situation of mounting socioecological problems, the failures of the state are scarcely recognised as limitations of the state per se. According to the Hegelian argument, the modern state cannot be wrong in itself and its mistakes are circumstantial imperfections that have rational correction. The Hegelian state system, with its malleability and operational precision, is a prefigured entity, something that is right by definition and can only advance additional rightness. If the state is never in the wrong, when a problem is detected it is either something in the world that must be out of place, or the understanding of the world is still imperfect and additional techno-science could close the gap. But the most perverse thing about the Hegelian mystification is that it provided renewed justification for the anti-commons and anti-democratic tendencies of the capitalist state. The Hegelian state is supposed to put into practice a 'police' approach that combines rationality, profitable market transactions and the moderation of conflicts derived from social inequalities. Hegel situates the state very close to God, it actually becomes part of God's project, but in practice it can only evolve according to the structures of the existing (capitalist) society. Following Hegel's political framework, rights and morality are associated with those that control the state (basically, the corporations and those who own private property): "at its highest point the political constitution is the constitution of private property. The highest political inclination is the inclination of private property" (Marx, 1970: 99). As a result, Hegel's clever rationalisation and idealisation of the state anticipated the more recent configuration of environmental statehood and the pursuit of the flexible interventions needed to create novel institutional arrangements for the circulation and accumulation of capital.

In his analysis of the political situation in France during the restoration of the monarchy, Marx denounced the opportunistic behaviour and the fluid alliances between the different sectors of the French elite, who saw in King Louis Bonaparte an adequate leader to secure 'bourgeois order' in the country. It is in that context that he famously observed that, while Hegel said that historical facts occur twice, he forgot to add that history repeats itself first as tragedy and then as farce (Marx, 1913). The evolution of environmental statehood, which has been systematically contained by the anti-commons imperatives of capitalist production and the political constraints of the contemporary state, also vividly reveals this combination of tragedy and farce. In the case of France in the middle of the nineteenth century, by protecting property relations and giving special treatment to the demands of industrial and commercial sectors, the new autocratic king strengthened the historical role of the bourgeoisie and helped to gradually consolidate the capitalist modernisation of the country. Marx's acute eyes detected the travesty of enthusiastic discourses endorsing the desired politico-economic reforms while the regime maintained a slavish adherence to old routines. The consequences of such political manoeuvres were the delights of *La Belle Époque* for the affluent members of the French society, the deepening of the conflicts around colonisation and economic liberalism and, ultimately, the horrors of the First World War. Paris was splendidly renovated by Baron Haussmann, not to serve the majority of its people, but to represent the prosperity and snobbery of the national elites. Obviously, any comparison between centuries and countries is anachronistic and problematic. Nonetheless, it is possible to identify parallels between the elitist basis of the environmental responses formulated by the state and the chicanery that brought the second Bonaparte to the French throne.

The recognition of the ambiguous character of flexible environmental statehood permits us to demystify the ordinary calls for both ecological modernisation and for the ecologisation of capitalist relations of production and reproduction. In recent decades – as part of the expansion of flexible environmental statehood – more national states have been encouraged to wield a 'green' rhetoric and demonstrate their socioecological commitments than was the case before during the prevalence of conventional environmental statehood. The state is expected to transform itself into something like an 'eco-state' able to promote a new kind of democracy that includes not only all the existing human beings, but also future generations and non-human beings (Eckersley, 2004). The eco-state is supposed to give more attention to its own environmental image and develop new abilities for dealing with socioecological issues. It is a type of state that attaches an even greater importance to the proper organisation and functioning of its own environmental statehood. The eco-state is likewise seen as an evolution of the welfare and of the neoliberal political formations because of its alleged ecological sensitivity towards the non-human realm of socionature (Barry and Eckersley, 2005). The archetype of the eco-state clearly incorporates elements of post-neoliberalism, as well as an improved discourse of sustainability and governance in an attempt to embrace the 'more-than-human' spheres of socionature. In practice, however, the advocacy of the eco-state suffers from the same faults and idealisation that undermines flexible environmental statehood. Although the supporters of the eco-state claim to have removed the false dichotomy between conventional (i.e. 'destructive') and responsible (i.e. 'green') states, this reconciliation operates only at the scale of meanings and functions instead of addressing the more fundamental political commitments and socioecological constraints of the contemporary (capitalist) state.

The argument in favour of an eco-state presented by Eckersley and her colleagues is unequivocally influenced by the sociopolitical ideas put forward by Habermas. As is well known, Habermas is one of the most important exponents of the so-called critical theory, that is, the reinterpretation of the connections between human agency and social structures in capitalist circumstances. Societies are described as complex formations that simultaneously comprise the sphere of systems and the sphere of the lifeworld; as a result, the main "problem of social theory is how to connect in a satisfactory way the two conceptual strategies indicated by the notions of 'system' and 'lifeworld'" (Habermas, 1987: 151). This gap between lifeworld and systems has major consequences not only in terms of interpreting the world, but also how to act upon it. The result of a growing, but problematically articulated, rationalisation of the lifeworld and its systemic differentiation (Habermas, 1987) has been an attempt by politicians and the broader political system to impose ideologies of social inclusion, for example through the mass media and the culture industry, without substantively removing social antagonisms and irrationalities (Habermas, 1991). The social policies of the contemporary state are correctly criticised by Habermas for the contradiction between trying to secure freedom on the one hand and cancelling it on the other. The Habermasian alternative to such shortcomings of state action is the pursuit of some common understanding and communication. Democracy should be, thus, enacted as collective deliberation or discursive democracy, which can be properly institutionalised in both the state and in interpersonal relations.

According to Eckersley (2004), despite some deficiencies in the communicative solution advanced by Habermas (she is particularly critical of him for not expanding his framework in order to embrace environmental justice and reflexive ecological modernisation), this politico-philosophical framework can inform the transformation of the current, destructive state into a 'green democratic state' (i.e. the eco-state). In the latter case, the focus is no longer on environmental statehood (as a sectoral response), but on the more general reconfiguration of the state as a truly ecological organism. Based on Habermas, spontaneous, grassroots initiatives should be able to both utilise and radicalise communication strategies. This would have important consequences for the renovation of the political system, since communicative action requires a democratic legal framework for its realisation. Moving along those lines, a more radical transformation could be achieved, resulting in the 'green state', which is supposed to be capable of promoting social equality and socionatural interaction. However, the Habermasian eco-state proposition fails to resolve the contradictions of an environmental statehood that remains subordinate to the stronger politico-economic commitments of the state. Despite the arguments in favour of more comprehensive forms of communication, an apparent democracy between the human and the non-human is never going to be secured if economic production and social reproduction remain contained by anti-commons imperatives. The main problem is not how the state deals with immediate environmental policies and strategies, but what the ultimate goals of the capitalist state are and how these are related to other processes of exclusion and exploitation.

Although Eckersley correctly criticises the anthropocentrism of Habermas's communicative ethics and the absence of nature in his political elaboration, she hesitates to properly identify the politicised interconnections between state, society and socionature. Eckersley and others closely follow Habermas's suggestion that the crisis of legitimacy that derives from the process of political struggle can be resolved through a radical strategy of communication. The Habermasian democratisation plan requires a rationalisation of public and private procedures in a way that depends on the (unproven) positive contribution of ample communication. His type of communication between speaking and acting subjects means something profoundly different to a radical and effective transformation of the state. Habermas's political philosophy deals with the two sources of legitimation – the rule of law and individual rights – which are expected to be reconciled through discourse and in the dimension of historical time (Habermas, 2001). The way forward is to consolidate the constitutional democracy and the democratic regime of law that guarantee a voluntary association of free and equal citizens (Habermas, 1996). However, there is scarcely any indication in the Habermasian political framework of how this strategy can lead to an effective democratisation of the state, let alone convert it into a more ecological influence as defended by the supporters of the eco-state. As mentioned several times already in this book, in a society with marked social asymmetries, the enrichment of communicative approaches cannot avoid being affected by the priorities of the stronger social groups and their perverse influence on the state. Environmental problems are neither going to be solved only by communication happening within the state, nor can communication be sufficient to contain and remove the perverse impacts of the state acting with capital accumulation priorities in mind.

The idealisation of the eco-state, informed by Habermas, is not much different to the framework of flexible environmental statehood inspired by Hegel's political project. In fact, the elements of the eco-state replicate the Hegelian plan for an enlarged democracy – in the case of Hegel, to include the peasantry, and for the eco-state, to include the rest of socionature – but still from the perspective of the existing relations of production and the preservation of social hierarchies. In other words, the defence of the eco-state fails to resolve the long-lasting inconsistency between environmental conservation goals and the superficial basis of democratisation. The simple appendage of an environmental dimension to the existing state apparatus – via an expansion of Habermasian democracy or via Hegelian-inspired environmental statehood – is certainly not adequate to alter the main economic and political responsibilities of the state. Equally, the mere advocacy of a 'strong ecological modernisation', to be adopted with an active oppositional public sphere (as suggested by Dryzek *et al.*, 2002), is not enough to bring environmental values to the attention of the state. The simple amelioration of the environmental agenda of the state produces only reactive, tardy and bureaucratic responses to socioecological questions that are deeply ingrained in the economic and political fabric of a mass consumption and mass waste society. The unavoidable compulsion of capitalism to expand into the non-capitalist domains (such as through the commodification of ecosystem services and monetisation of environmental conservation) is nothing less than a 'death drive' [*Todestrieb*] that resembles the Freudian description of the life instinct that coexists with the primitive impulse for destruction, decay and death.

The insufficiencies of the mainstream case for the green state, and the limited efforts to go beyond the existing socioeconomic relations, betray the enduring influence of the European Enlightenment.[2] The political thinking of the time desperately attempted to justify the separation between civil society and political society and set limits on the interference of the state apparatus in the realm of capitalist interests (this is demonstrated notably in the work of both Kant and Hegel). It is precisely this ideological divorce between the social and the political spheres of life that sows the seeds of the invalidation of environmental statehood. Posited as an arbiter of society, the state is nonetheless unable to detach itself from social and ecological disputes, and ends up favouring mainly the stronger economic and political interests. Following the dichotomic premises of the Enlightenment, the root causes of environmental problems (i.e. the unyielding pursuit of capital from the double exploitation of society and socionature) are never open for questioning. Environmental problems are left in the hands of the state to be resolved almost exclusively through technical and legal approaches that unavoidably preserve the mechanisms of private capital accumulation and the political influence of hegemonic groups (even if isolated processes of capital accumulation are circumstantially affected by environmental regulation, fees or penalties). The history of the last century proved that environmental degradation has been initially promoted and then appropriated by the state through initiatives that facilitate the access to, and the control and exploitation of socionatural systems and resources. For those who benefit from the exploitation of society and socionature, the political system should not impinge on fundamental economic liberties, and, consequently, environmental statehood is only allowed to operate reactively and after the environmental impacts are already established.

Therefore – from a critical political ecological perspective – what is required is a radical and coordinated abolition of the narrow models of environmental statehood that have achieved, at best, only a temporary mitigation of impacts and a superficial response to socioecological demands. With the deconstruction and removal of the Hegelian basis of flexible environmental statehood (i.e. the mystification of problems and the stratification of social engagement), what should emerge is the possibility of addressing the politicised basis of environmental problems and the ecological features of state action. This entails a much broader process of change that profoundly connects sectoral environmental politics with transformations in the other socioeconomic areas overseen by the state. As contended by Marx, "social struggle was everywhere, fought between the *dominated* and the *dominant*, and this *within* the state, *within* civil society and by extension *within* anything else in society" (Goonewardena and Rankin, 2004: 138). The important conclusion to be drawn here is that a genuine democratisation of the state is a prerequisite for the resolution of environmental questions, and effective solutions to persistent environmental problems play a key role in the improvement of democracy and statecraft.

The central responsibility of those concerned with the political ecology of the state – both in the academic and non-academic arenas – is to expand the critique of the existing paradigm of environmental statehood and search for emancipatory mechanisms of environmental regulation, conservation and management. It is the rationale and the commitments behind the environmental responses by the state apparatus that need to be, first and foremost, questioned and reoriented. As Marx (2012: 44) pointed out, "the working class cannot simply lay hold of the ready-made state machinery", but this machinery needs to be profoundly transformed, just as much as the economy and society need to substantially change. It is not enough to merely free the state from public interferences; the public expects a conversion of the state from an organ superimposed upon society into one completely subordinate to it (Marx, 2001). Marx (1975) also advised that it is the very manner of emancipation that needs to be criticised, given that the state can liberate itself – from religion, in the case of Marx's analysis, and from the burdens of environmental degradation in the case of political ecology – without people being set free as well. The political emancipation of the religious person was dialectically located by Marx in the wider process of liberation of the state from religion in general and the pursuit of full human emancipation. The persistence of a condition of economic subalternity is closely connected with political alienation, which is a distortion of democracy produced by the most powerful groups and whose source "must be looked for in the *nature* of the state itself" (Marx, 1975: 217).

This challenging plan of action is certainly even more difficult to implement now than a few decades ago due to the highly adverse political climate and the limited opportunities to challenge established, neoliberalised practices. However, despite the political adversity faced by those holding critical views, the recognition of the political ecology of the state should remain an integral part of the strengthening of democracy and the removal of socioecologically degrading pressures. The ability of civil society to govern itself without the constraints imposed on it by the state is something that Gramsci (1975: 662) described as "*il riassorbimento della società civile nella società politica*" [i.e. the collapse or the reduction of civil society in political society]. This observation has many practical consequences, particularly in Western European countries, where for several decades environmental questions have been dealt with, with passivity and alienation. Against the background of those perverse disorganising trends, the resistance to the authoritarianism of conventional socioeconomic policies must be achieved with alternative epistemes and procedures, which should operate as a legitimate statecraft from the bottom up that could open up "the way for alternatives of many kinds" (Hecht, 2011: 214). The necessary transformation of the state must be seen as a central element of the emancipation of socionature from political domination, ideological hegemony and, ultimately, socioecological exploitation. It means gradually phasing out the existing forms of state and the construction of a democratic society in which the state, at least in the format we know it by, is ultimately less necessary (Marx and Engels, 1974). In the words of Lefebvre,

> the more the functions of State power are exercised by the whole of the people, the less necessary this power becomes. This is what Lenin himself calls the revolutionary dialectic of Marx. The theory of the State at the end of the State and, more generally still, the political theory of Marx aims at the end of all politics.
>
> (Lefebvre, 2009: 88)

Finally, a transformation of the state that really contributes to a fairer, more democratic society and removes the exploitation of socionature should happen according to solid ethical practices consistent with eco-socialist values. It is still certainly the case that ethics and values remain among the most delicate topics for those with socialist inclinations. But a proper political ecology cannot ignore the importance of the moral principles affecting everyday decisions and long-term strategies. The fundamental modification of the state requires changes at a personal and intersubjective level, as well as changes to the wider rationalities and commitments of the state. As theorised by Freire (1996: 61) on the subject of revolutionary education and social justice, "people teach each other, mediated by the world". And when facing the difficult reality of the early twentieth century, another South American author, José Mariátegui, thought that there was no prospect of a better world without the recognition of the rights of those historically excluded and the consideration of unconventional issues, such as race, gender, culture and nation, which were ignored by the critical political opposition. The true transformation of the state and the nation can only be achieved according to clear ethical principles – what Mariátegui (2011) called the 'ethic of the proletariat' – and must be expanded today into a sort of 'politico-ecological ethic'. In contrast to conventional moralist ideas, the ethics of change are a key element of the elimination of long-lasting injustices and the removal of the socioecological trends maintained, first and foremost, by the state.

In the near future, environmental statehood is likely to become an especially controversial arena fraught with conflicting interests and disputes. Those clashes will continue to be a result of the escalation of anti-commons policies and the pro-commons resistance of groups and communities excluded from the prevailing patterns of development. Addressing the challenges ahead will require a clear grasp of the significance of class identity for the eradication of the double exploitation of society and of the rest of socionature. The liberation of marginalised classes from environmental degradation must be connected with the elimination of the anti-commons allegiances of the state and the suspension of the socioecological contradictions of capitalist relations of production and reproduction. As Marx convincingly argued in his infamous court trial in Cologne in 1849, "you cannot make the old laws the basis of new society" (in Mehring, 1936: 181). Thinking in terms of class-struggle constitutes a real force for change and produces the opportunity for collective learning and social transformation, including the radical reorganisation of the state. The fight for the political recognition of the subaltern classes and groups is directly connected with the reconciliation of the social and natural dimensions of socionature, (long) separated because of the expansion of the capitalist relations of production and reproduction. Likewise, the transformation of the state is a ramification of the wider processes of class struggle, given that clashes between classes are, by definition, battles fought 'in and against' the state (London Edinburgh Weekend Return Group, 1980). It is evident, though, that in a globalised and highly complex world, "class struggle has to be refocused as a struggle against [unjust] class relations, without the comfort of a possible universal class emerging" (Sitton, 1996: 250). But it is in that context of confusing class identities that a critical politico-ecological agenda can help to catalyse efforts towards political emancipation and the resolution of socioecological tensions. There is still a long way to go and a huge amount to learn, but the recent and ongoing crises, since 2008, have had at least one positive outcome, as highlighted by Hobsbawm (2011: 417): "something has changed for the better. We have rediscovered that capitalism is not the answer, but the question."

Note

1 Needless to say, non-capitalist societies have their own political and social problems, including processes of exploitation and domination. The difference, however, is that a capitalist order is organised around the systematic appropriation of territorial resources, the generalisation of exchange-value and the maximisation of private capital accumulation.
2 Notwithstanding all the positive aspects of the Enlightenment (as discussed, for example, by Callinicos, 2007).

Bibliography

Avineri, S. 1973. Labour, Alienation, and Social Classes in Hegel's *Realphilosophie*. In: *The Legacy of Hegel*, O'Malley, J.J., Algozin, K.W., Kainz, H.P. and Rice, L.C. (eds). Martinus Nijhoff: The Hague, pp. 196–215.
Barry, J. and Eckersley, R. (eds). 2005. *The State and the Global Ecological Crisis*. MIT Press: Cambridge.

Callinicos, A. 2007. *Social Theory: A Historical Introduction*. Polity: Cambridge.

Cullen, B. 1979. *Hegel's Social and Political Thought: An Introduction*. Gill and Macmillan: Dublin.

Dryzek, J.S., Hunold, C., Scholsberg, D., Downes, D. and Hernes, H.-K. 2002. Environmental Transformation of the State: The USA, Norway, Germany and the UK. *Political Studies*, 50, 659–82.

Eckersley, R. 2004. *The Green State: Rethinking Democracy and Sovereignty*. MIT Press: Cambridge, MA and London.

Freire, P. 1996. *Pedagogy of the Oppressed*. Trans. M.B. Ramos. Penguin: London.

Goonewardena, K. and Rankin, K.N. 2004. The Desire Called Civil Society: A Contribution to the Critique of a Bourgeois Category. *Planning Theory*, 3(2), 117–49.

Gramsci, A. 1975. *Quaderni del Carcere*. Einaudi: Torino.

Habermas, J. 1984. *The Theory of Communicative Action: Reason and Rationalization of Society*. Vol. 1. Trans. T. McCarthy. Polity Press: London.

Habermas, J. 1987. *The Theory of Communicative Action: The Critique of Functionalist Reason*. Vol. 2. Trans. T. McCarthy. Polity Press: London.

Habermas, J. 1991. *The Structural Transformations of the Public Sphere: An Inquiry into a Category of Bourgeois Society*. Trans. T. Burger. MIT Press: Cambridge, MA.

Habermas, J. 1996. *Between Facts and Norms: Contribution to a Discourse Theory of Law and Democracy*. Trans. W. Rehg. MIT Press: Cambridge, MA.

Habermas, J. 2001. Constitutional Democracy: A Paradoxical Union of Contradictory Principles? *Political Theory*, 29(6), 766–81.

Hecht, S. B. 2011. The New Amazon Geographies: Insurgent Citizenship, 'Amazon Nation' and the Politics of Environmentalism. *Journal of Cultural Geography*, 28(1), 203–23.

Hobsbawm, E. 2011. *How to Change the World*. Abacus: London.

Lefebvre, H. 2009. *State, Space, World: Selected Essays*. Trans. Moore, G., Brenner, N. and Elden, S. University of Minnesota Press: Minneapolis, MN.

London Edinburgh Weekend Return Group. 1980. *In and Against the State*. Pluto Press: London.

Luke, T.W. 1990. *Social Theory and Modernity: Critique, Dissent, and Revolution*. SAGE: London.

Mariátegui, J.C. 2011. *An Anthology*. Ed. and trans. H.E. Vanden and M. Becker. Monthly Review Press: New York.

Marx, K. 1913 [1852]. *The Eighteenth Brumaire of Louis Bonaparte*. Trans. D. de Leon. Charles H. Kerr & Company: Chicago.

Marx, K. 1970 [1843]. *Critique of Hegel's 'Philosophy of Right'*. Trans. A. Jolin and J. O'Malley. Cambridge University Press: Cambridge.

Marx, K. 1975. *Early Writings*. Trans. R. Livingstone and G. Benton. Penguin and New Left Review: London.

Marx, K. 2001 [1875]. *Critique of the Gotha Programme*. The Electric Book Company: London.

Marx, K. 2012 [1871]. *The Civil War in France*. Aristeus: Chicago.

Marx, K. and Engels, F. 1974 [1845–6]. *The German Ideology*. Trans. C.J. Arthur. Lawrence and Wishart: London.

Mehring, F. 1936. *Karl Marx: The Story of His Life*. Trans. E. Fitzgerald. John Lane: London.

Plant, P. 1973. *Hegel*. George Allen & Unwin: London.

Sitton, J.F. 1996. *Class Formation and Social Conflict in Contemporary Capitalism*. SUNY Press: Albany.

Index

Page numbers in *italics* denote tables.

Stewart, J. 143, 145
SUNASS (Superintendencia Nacional de
 Servicios de Saneamiento) 124, 128,
 130, 132
Swyngedouw, E. 31, 34

Taylor, C. 150
technocratic 92, 120, 167; heritage 119;
 perspective 128, 130, 135;
 policy-making 49, 108
techno-environmental 110; adjustments
 104–5, 127; dimension 103, 124;
 improvements 104–5, 134; initiatives
 105; results 105
Technological Revolution 44
Tennessee Valley Authority (TVA) 46–7
Thompson, E.P. 3, 17
trialectical 35–6, 167; interpretation 15;
 relationship 37, 39n2, 43
trialectics 35–7, 57, 63, 140

United Nations 9, 24, 105; Conference on
 Environment and Development 68;
 Dublin Principle 85; International
 Conference on Water and the
 Environment 85; International
 Drinking Water Supply and
 Sanitation Decade 104; Water
 Conference 103–4
United States 46, 69, 74–5, 161;
 environmental law 55; Environmental
 Protection Agency (EPA) 48;
 environmental statehood 48–9; Free
 Trade Agreement 127; Independence
 44; National Environmental Policy Act
 (NEPA) 48, 56, *71*; presidents 46, 73,
 94n1; USAID 128
urban 122–3; development 57, 127, 133,
 135; drainage 61; expansion 59;
 oligarchs 84; policies 126; policy-
 making 17; poor 121, 156; socionature
 120; water supply 58, 114, 134

urban–industrial 166; capitalism 11, 31,
 42, 46; expansion 45; relations of
 production and reproduction 44;
 socioeconomy 42
urbanisation 58, 66, 110, 120–1; neoliberal
 urbanism 122

Vargas Llosa, M. 27, 122

water 104; drinking 62, 88, 128;
 transference 85, 129
Water Framework Directive (WFD) *71*,
 109, 111, 120; charges 115;
 implementation in Scotland 112–19;
 national legislation 110–12; regime 110,
 114–15, 118; regulators 110, 117; River
 Basin Management Plans 113, 118
water management 46, 83–4, 91, 109–10,
 114, 120, 133; in Brazil 85;
 complexification 119; decisions 116;
 economic dimension 104; issues 105;
 key disciplines *131*; modernisation 159;
 problems 90, 92, 111, 113; reforms 93,
 128
water neoliberalisation 102–6, 127–30,
 134–6; in Lima 124, 128, 130, *131*,
 132–5
water sector 159; Brazilian 83, 89, 91, 93;
 Latin America 91; Scottish 111;
 transformations 106
welfare-developmentalist 50, 66, 105;
 policies 62; state 6, 16, 20n5, 63, 68,
 139, 167; strategies 72
Wells, D. 9, 43
West, D. 144
Whitehead, M. 5, 14, 29
Williams, H. 76, 144
Wood, A.W. 158
World Bank 9, 61–2, 104–5, 123–4, 128
Wurzel, R.K.W. 49, 69

Zizek, S. 147–9